Physical Processes in Comets,
Stars and Active Galaxies

Physical Processes in Comets, Stars and Active Galaxies

Proceedings of a Workshop, Held at Ringberg Castle, Tegernsee, May 26–27, 1986

Editors: W. Hillebrandt,
E. Meyer-Hofmeister, and H.-C. Thomas

With 46 Figures

Springer-Verlag Berlin Heidelberg New York
London Paris Tokyo

Wolfgang Hillebrandt
Emmi Meyer-Hofmeister
Hans-Christoph Thomas

Max-Planck-Institut für Physik und Astrophysik,
Institut für Astrophysik, Karl-Schwarzschild-Str. 1,
D-8046 Garching, Fed. Rep. of Germany

ISBN 3-540-17766-3 Springer-Verlag Berlin Heidelberg New York
ISBN 0-387-17766-3 Springer-Verlag New York Berlin Heidelberg

Offset printing: Weihert-Druck GmbH, D-6100 Darmstadt
Bookbinding: J. Schäffer GmbH & Co. KG., D-6718 Grünstadt
2153/3150-543210

Preface

In May 1986 a two-day workshop on Physical Processes in Comets, Stars and Active Galaxies was held at the Ringberg Castle near Lake Tegernsee, and this rather unusual collection of topics needs a few words of explanation.

When we first thought of organizing a workshop on such a large variety of astrophysical objects our main motivation was to honor Rudolf Kippenhahn and Hermann Ulrich Schmidt on the occasion of their 60th birthdays, and we planned to cover at least a fraction of their fields of active research. We then realized immediately that despite the fact that the objects are so different, the physical processes involved are very much the same, and that it is this aspect of astrophysics which governed the scientific lives of both of our distinguished colleagues and friends and allowed them to make major contributions to all those fields. Apparently this viewpoint was shared by many colleagues and it was therefore not surprising that in response to our invitation everybody who had been invited agreed to come and to present a talk.

The workshop then turned out to be a real success. In contrast to highly specialized conferences, fundamental problems as well as very recent developments were discussed and the participants appreciated the opportunity to exchange ideas. Since all talks were given by experts in the various fields we thought it might be worthwhile and of general interest to publish their excellent contributions to the workshop in a Festschrift, thereby expressing our gratitude and admiration to Rudolf Kippenhahn and Hermann Ulrich Schmidt who over many years have shown us how exciting science can be. The book, therefore, not only reflects the present status of our understanding of a variety of astrophysical objects but is also meant to be a small sign of gratitude for all their efforts in promoting astrophysical research.

We would like to thank all speakers for contributing written versions of their talks and for helping us to edit the proceedings of such a unique conference.

Garching, January 1987

W. Hillebrandt
E. Meyer-Hofmeister
H.-C. Thomas

Contents

Part IV **Active Galactic Nuclei**

Part I

The Physics of Comets

Comet Explorations

W.F. Huebner

T-4, Los Alamos National Laboratory, Los Alamos, NM 87545, USA

Several new missions to short period comets have been proposed. It is expected that these missions will extend our knowledge not only about comets, but also about the chemical and physical properties of the nebular region from which comets originate. Contrary to what is often implied, cometary material is not the pristine material of the nebula in which cometesimals formed. The gases in the coma have been chemically and physically processed by solar radiation and solar wind. The icy components of the nucleus have been altered by cosmic rays, ultraviolet radiation, the solar wind, and by radioactivity during the several 10^9 years that a comet spends in the Oort cloud. Even during the formation processes of cometesimals, selective condensation and chemical reactions have destroyed the chemical or physical records from the nebular region in which comet nuclei originated. To obtain the chemical composition and physical properties of the original nebular region from in situ measurements on comets, computer modeling, supplemented by laboratory determinations of material properties, is required. This general objective of modeling with laboratory support should be part of all mission plans.

1. INTRODUCTION

Several new comet missions, among them the Comet Rendezvous and Asteroid Flyby (CRAF) mission, the Multi-Comet (MC) mission, the Comet Atmosphere Encounter and Sample Return (CAESAR) mission, and the Comet Nucleus Sample Return (CNSR) mission to short period comets have been proposed for the decades ahead. These missions are NASA, ESA, or joint NASA and ESA missions. It is expected that they will extend our knowledge not only about comets, but also about the chemical and physical properties of the nebular region in which comets formed. This region may be the solar nebula, the presolar nebula, or a companion fragment of the presolar nebula.

Contrary to what is often implied, cometary material is not pristine nebular material; "new" comets are pristine only in the sense that they have never entered the inner solar system since their formation. That the gases in the coma have been chemically and physically processed has long been recognized. But even the icy

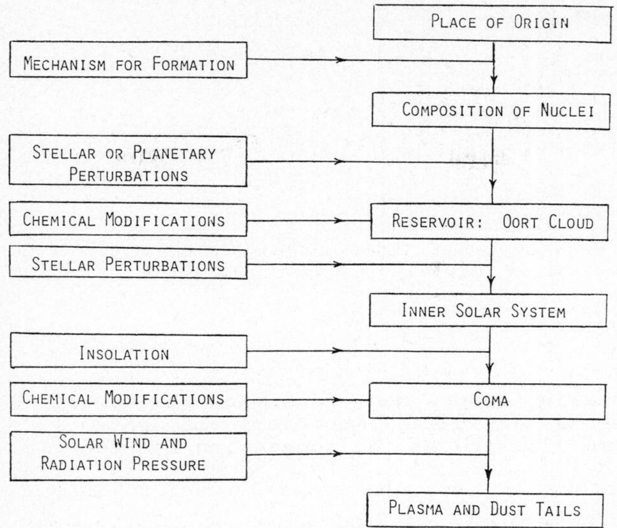

Fig. 1: Block diagram illustrating the history of processes that influence the physical and chemical structure of comets.

component of the nucleus has been processed by cosmic rays, ultraviolet radiation, the solar wind, and by internal radioactivity during the several 10^9 years that a comet spends in the Oort cloud or in the inner comet belt. Still earlier, during the formation process of a cometesimal, selective condensation and chemical reactions have destroyed any direct chemical or physical similarity between the chemical abundances in the nebular region and a nascent comet nucleus.

To obtain the chemical composition and physical properties of the nebular region in which comets formed from in situ measurements made during spacecraft explorations of comets, computer modeling, supplemented by laboratory determinations of material properties, is required. Figure 1 illustrates the mechanisms (on the left) that act on long period comets from their place and time of origin until the observations can be made in the inner solar system. All in situ measurements will be made on short period comets, i.e., on comets that have past through the inner solar system several times. Long period comets would be more ideal for the proposed measurements, but their orbits cannot be determined accurately enough, early during the approach of a new comet into the inner solar system, so that a spacecraft could be directed to intercept them. Thus modeling assumes the additional significance of interpreting and understanding the data collected from short period comets and transposing and projecting this new knowledge to long period comets.

Figure 2 illustrates the interrelationship between (a) ground based and in situ observations, (b) computer models, and (c) experiments to obtain needed data as well as experiments to simulate cometary con-

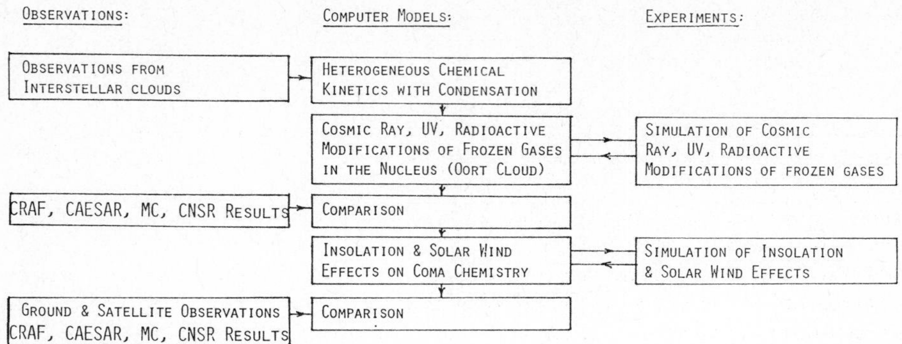

Fig. 2: Block diagram for modeling comets and connecting the model with observations and experiments, including laboratory and space simulations. This diagram also illustrates the connection between space science and astrophysics.

ditions. By necessity of a simple block diagram, many of the intermediate steps are not indicated. For example, in the top, central box "condensation" also implies formation of cometesimals under the action of accumulation and aggregation. Comparing figure 2 with figure 1 helps to understand and fill in many of the intermediate steps.

In summary, the problems are:

1. The place of origin and the mechanism of formation of cometesimals are not known.
2. Chemical models of dense interstellar clouds are incomplete.
3. Chemical modifications of comet nuclei in the Oort cloud or in the inner comet belt are just beginning to be investigated.
4. Physico-chemical models of comet comae are incomplete.
5. All in situ measurements are on short period comets.
6. All measurements on long period comets are by remote sensing and therefore also incomplete.

Of particular importance are the modeling and laboratory simulation efforts that are relevant to item (3), which has received very little attention, and that are in support of the activities relevant to item (5). Item (3) in the above list also corresponds to the boxes in the second row of figure 2. This research will be useful (a) to develop models for the chemical modification of materials in comet nuclei in the Oort cloud or the inner comet belt, (b) to obtain experimental data in the laboratory for dusty ice mixtures relevant for short period comets or for comets in orbits with small perihelion distances coming from the Oort cloud, and (c) to simulate in the laboratory certain processes occurring in comets. We believe that this research, when coupled to the proposed comet missions, will be an important step to extend our knowledge in the exploration of space beyond the solar system into the surrounding galaxy. It provides the bridge from space science to astrophysics.

2. DEVELOPMENT OF A MODEL TO INVESTIGATE NEBULA PROPERTIES THAT LEAD TO COMET FORMATION

Traditionally it is thought that comets enter the inner solar system for the first time because passing stars perturb their orbits in the Oort cloud and short period comets (P < 200 years) are considered the result of dynamical evolution of "new" or long period comets that have perihelion distances close to the orbits of Jupiter or Saturn. But it has also been suggested that short period comets come from an inner comet belt that exists beyond the region of the outer planets (Everhard, 1972). Thus several options for the physical and chemical properties and thermodynamic conditions must be considered.

There are two major hypotheses about the origin of comets. In the first hypothesis, comets were formed at the same time as the sun and the planets. After fragmentation and collapse of a molecular cloud, planets formed out of the gas and dust in the rotating disk that surrounds the proto-sun. Small bodies that did not accrete into planets remained as cometesimals. In this scenario cometesimals are formed in the outer planetary system or just beyond it. According to the second hypothesis, cometesimals are aggregates of ice-covered interstellar dust. In this case they are formed in parts of the presolar nebula that have not been reprocessed during formation of the solar system. To investigate both hypotheses, models appropriate for the solar nebula (SN), the presolar nebula (PSN), or a companion fragment of the PSN must be considered. In one of the extreme conditions, at the highest appropriate densities and temperatures, the chemical composition will be closer to equilibrium than in the other cases and the dynamic effects may be dominated by turbulence. In the other extreme will be time - dependent heterogeneous chemical kinetics with dynamical evolution dominated by damped oscillatory motion through the plane of rotation of the disk (sedimentation). Conditions akin to a chemical steady state may lie in between. In all cases the interplay between chemical and dynamical evolution with condensation and accumulation on grains must be investigated, leading to aggregation and formation of small bodies of icy - dust mixtures. Figure 3 illustrates possible choices and formation mechanisms.

The difficulty with present models of interstellar clouds with condensation is that the condensation of heavy species on grains is so efficient and fast that the cloud is depleted of heavy molecules in several 10^4 years. This seems to contradict observations. Abundances of heavy molecules, much higher than predicted by models, are observed in interstellar clouds that are believed to be much older than 10^5 years. Processes that return heavy molecules to the gas phase appear to be missing. New mechanisms must be considered that increase the abundances of heavy molecules in the gas phase. One such possible mechanism is that electrons, since they are much more mobile than

Place of Origin	Physico-Chemical Mechanisms in Formation
Companion to PSN or Outer PSN	Heterogeneous Chemical Kinetics with Condensation and Dynamics
Intermediate PSN	Chemical Steady State with Condensation and Turbulence
Outer Planetary Region	Near Chemical Equilibrium with Condensation

Fig. 3: Relationship between place of origin and formation mechanism.

ions, charge grains negatively when they collide with them. The negatively charged grains will then have a much larger Coulomb cross section for collisions with positive ions. If some condensation or at least a monolayer of molecules has formed on these grains, then a positive ion interacting with a grain can simultaneously react with the molecules on the surface. An electron recombination occurs spontaneously. In some of these surface reactions larger molecules will be formed. The possible formation mechanism of molecules on neutral grains has been realized for some time, but there is not enough energy to get the newly formed heavy molecule off the grain into the gas phase. However, with the electron recombination occurring simultaneously on charged grains, there will be enough energy liberated to "vaporize" several molecules from the grain surface into the gas phase. This scenario has not been investigated in detail.

Heterogeneous chemical reactions may play an important role. Silicates may be good catalysts for the formation of some hydrocarbons including CH_4 and C_2H_4. These processes require more basic research before they can be incorporated into comprehensive comet models.

Techniques to model nucleation with subsequent condensation for grain formation are at the threshold of new developments. Gail and Sedlmayr (1984) and Gail et al. (1984) have developed such techniques for carbon grain formation.

Development of comprehensive comet models is a long range program that needs to go through several phases, not only for the various conditions and processes that must be investigated as is apparent from figure 3, but also from the point of view of incorporating newly developed concepts for the basic physical processes. Such a project might be divided into the following specific tasks:

1. Develop a model for the intermediate solar nebula with chemical steady state (nearly chemical equilibrium), condensation, and turbulent - like motion.

2. Develop a model for the outer solar nebula with gas - phase chemical kinetics (nearly in chemical steady state), condensation, and dynamical evolution.
3. Develop a model for the PSN with gas - phase chemical kinetics, condensation, and dynamical evolution (sedimentation).
4. Develop a model for a companion fragment of the PSN with gas - phase chemical kinetics, condensation, and dynamical evolution (sedimentation).
5. Supplement the gas - phase chemical kinetics with heterogeneous chemical reactions.

These tasks are listed in order of complexity, always building on the model and the results from the previous task. Some specific features of these tasks are:

1. The model for the intermediate SN concerns itself with the space now occupied by the outer planets and somewhat beyond. The density and temperature in this region are such that near chemical equilibrium may exist. By this is meant that H_2O, CH_4, and NH_3 are the major chemical constituents with dust and CO, CO_2, HCN, and C_2H_2 are important, but minor species and most other chemical species are present only as traces. The detailed chemical composition will vary with heliocentric distance and the relative abundances need to be determined in detail. Turbulent mixing will influence the relative abundances. Because of the relatively high density in this region, condensation, accumulation, and aggregation (sweep-up) into cometesimals may occur very rapidly. The disadvantage is that the cometesimals that might be formed are in the ecliptic and only planetary perturbations can get them into the spherical Oort cloud (Remy and Mignard, 1985; Mignard and Remy, 1985). It is easier to get them into an inner comet belt. Computer programs for modeling chemical equilibrium with condensation (statistical) are available. This should not imply that these equilibrium conditions are valid for this region, but it permits a rapid examination about the relevance of the various physical processes. In the long term, the chemical equilibrium must be replaced by gas-phase chemical kinetics.

2. The model for the outer SN encompasses the space outside the region considered by model 1. In the inner part of this region the density and temperature are such that a chemical steady state may be a reasonable first approximation. Dynamic mixing will influence the relative abundances which need to be calculated using gas-phase chemical kinetics and gas - dust collision frequencies with sticking coefficients in a detailed model. In this case the densities are lower than in model 1 and cometesimal formation is slower. Radiation pressure may play an important role in the cometesimal formation (Hills, 1982; Hills and Sandford, 1983a,b). Stellar perturbations can change the orbits to get the cometesimals into the Oort cloud (Remy and Mignard, 1985; Mignard and Remy, 1985).

3. In the model for the PSN, simulation of the dynamics for the collapsing cloud must be included. Existing models (Tscharnuter, 1980, and references therein) can be parameterized. Gas-phase chemical kinetics and gas - dust collision frequencies with sticking coefficients to calculate monolayers, "condensation", and formation of ice mantles on grains can be taken over from the final version of model 2. Fluid dynamic instabilities and radiation pressure may play an important role in the formation of cometesimals.

4. The model for the companion fragment of the PSN has many similarities with model 3. The companion fragment is a smaller cloud that has separated from the original cloud. It is not big enough to form a star, but it can form cometesimals. The physical processes for cometesimal formation have been investigated by Biermann and Michel (1978). This model should be investigated and developed further by including the gas-phase chemical kinetics and detailed "condensation" on grains. It has the appealing features that stellar perturbations can easily extract cometesimals from the companion fragment. The small number of these cometesimals that will be captured by the main cloud of the PSN will have a random (isotropic) distribution similar to that of the Oort cloud.

5. The above four models need to be improved through developing and incorporating heterogeneous chemical reactions into the chemistry network. Formation of monolayers on grains will be an important process. Associated with this may be the production of heavy molecules in the gas phase. Present models underestimate production of heavy gas-phase molecules.

3. DEVELOPMENT OF A MODEL FOR THE STRUCTURE AND EVOLUTION OF COMET NUCLEI

After comet nuclei have been formed and transported into the Oort cloud (a spherical distribution out to a distance of several 10^4 AU from the Sun), but before they are perturbed by passing stars that will bring them either into the inner solar system or eject them from the solar system, they are thought to be in "storage". This leads to the impression that their physical and chemical state remains unaltered from the original conditions under which they were formed. In fact, inherent radioactivity of isotopes in the comet nucleus, cosmic radiation, ultraviolet radiation, interstellar winds, and a very weak but long lasting solar wind will alter their composition to varying depths below their surface. Very little modeling has been reported for chemical and physical changes occurring in comet nuclei while they are in storage in the Oort cloud or the inner comet belt. Hypotheses have been advanced that extraterrestrial organic compounds have been produced on interstellar dust grains and on the surfaces of comets in the Oort cloud (Greenberg, 1982). Chemical modifications resulting from ionizing and dissociating cosmic radiation (Draganic et al., 1984) and from radiochemistry, including radioactive heating (Wallis, 1980), have been investigated qualitatively. Experiments on pure ices at low temperatures have been linked to comets in the Oort cloud. These experiments are for pure water ice and for mixtures of frozen gases exposed to uv radiation, and for proton and heavier ion bombardment (Strazulla et al., 1983, 1984; Bar-Nun et al., 1985a) and are usually only applied to the evolution of comets as they approach the inner solar system. But these processes and the sputtering of ices by ion bombardment (Brown et al., 1978, 1980, 1982; Lanzerotti et al., 1978, 1982; Johnson et al., 1982, 1983a, 1983b) and the synthesis of heavier molecules and polymers from ion bombardment of frozen water, sulfur dioxide, and methane (Strazulla et al., 1984; Foti et al., 1984; de Vries et al., 1984; Lanzerotti et al., 1985) will clearly also apply to comets in the inner comet belt or in the Oort cloud (Pirronello, 1985). Many basic observations about the structure and origin of cometary nuclei have been made by Donn and Rahe (1982). However, no comprehensive model exists that links and combines all of the known information relevant to comet nuclei in the Oort cloud or the inner comet belt and simulates the physical and chemical modifications over time spans of several 10^9 years.

The thermal evolution of a pure ice nucleus has been modeled by Herman and Podolak (1985), Smoluchowski (1985), and Klinger (1985). Prialnik and Bar-Nun (1986) have modeled the evolution of a pure ice nucleus over many revolutions through the inner solar system. They find that transition from amorphous to crystalline ice does not occur regularly in every revolution, but only in a few revolutions over the life of a comet. How their findings would be changed if the comet had been warmed by radioactivity has not been investigated. Most comets have dust mixed with their icy components. This dust can change the heat capacity and the conductivity of the nucleus. A first model for a dust mantle and an ice - dust mixture has been developed by Podolak and Herman (1985). Effects of the dust properties on the structure and evolution of a comet nucleus will also vary with the quantity, size distribution, and the type of dust (carbonaceous or silicate). A dusty mantle on the surface of the nucleus will change the boundary condition for the temperature profile into the nucleus. The models of Herman and Podolak, Podolak and Herman, and Prialnik and Bar-Nun are important improvements over earlier models (e.g., Weissman and Kieffer, 1981, 1983; Rickman and Froeschle, 1982a,b; Kuehrt, 1984) in the analysis of thermal evolution of comet nuclei. The number of variables is very large: The dust size distribution, the composition of the dust, the ratio of dust to ice, and the composition of the ice. Chemical compositions for comets have been estimated by Delsemme (1982). Festou (1984) found in his analysis of many different comets that the composition of ice, i. e., the ratio of C to OH, varies greatly from comet to comet.

Much experience has been gained in chemical kinetics with associated time dependent physical changes, including energy balance, from modeling the coma of comets (for summaries see, e.g., Huebner, 1985; Huebner et al. 1986). These give excellent agreement with in situ comet measurements (Boice et al. 1986). The starting point for a nucleus model will be the assumption that the nucleus is composed of frozen gases, dominated by water in an amorphous state, mixed with varying amounts of dust. Radioactive heating of the mixture and exposure to cosmic radiation will be the primary modifying reactions for the bulk of the comet nucleus. There are many parallels that can be drawn between existing comet coma models and a comet nucleus model. For example, in the coma model one considers wavelength dependent attenuation coefficients, resulting in deposition of photons with different energies at different depths in the coma. In the nucleus model cosmic rays with different energies are deposited at different depths depending on the stopping power of the ice and dust. In the coma model, fast light particles, such as atomic and molecular hydrogen, carry kinetic energy relative to the escaping bulk gas. In a nucleus model gas molecules produced from radioactive heating, cosmic rays, or exothermic phase transitions of the ice will diffuse and carry energy with them. Chemical kinetics is an important process in the coma model. Ionizing radiation will produce similar reactions in the nucleus. The resulting structure and evolution of the nucleus must be obtained over a period of $4.5\ 10^9$ years, while the nucleus is in the Oort cloud or the inner comet belt. Energy balance, diffusion of gases, adsorption and absorption of gases by the dust grains and trapping of gases in the ice will have to be included in a comprehensive model. Ultraviolet radiation and solar wind effects will have to be included in the modeling of the surface. The effect of nuclear mutations induced by radionuclides in the comet nucleus and by cosmic radiation needs to be investigated.

Some of the important questions to be answered are: Is it possible to produce mixtures, for example varying ratios of dust to gas, that will warm up the nucleus to temperatures were amorphous ice can crystallize while the comet is in the Oort cloud or the inner comet belt? What fraction of trapped gases escapes from the nucleus? How is the chemical composition of the frozen gases altered? Can one expect gradients in the composition?

Certain rate coefficients and properties of ice - dust mixtures needed for modeling will not be known. The important processes for which data are lacking will have to be identified and measured as indicated in the next section.

4. LABORATORY DETERMINATION OF PROPERTIES OF ICE - DUST MIXTURES

Properties of ices have been investigated by many scientists at various research institutes. A conference on ices in the solar system was held in 1984 (Klinger et al. 1985). The importance of the exothermic transformation of amorphous ice into cubic ice as an internal energy source in comets has often been stressed by Patashnik et al. (1974). Smoluchowski (1981) and Klinger (1980, 1981) developed the idea further by considering that the phase transition will increase the thermal inertia of the nucleus. Amorphous water ice has been investigated by Mayer and Pletzer (1985). Experiments with gases trapped in amorphous ice at low temperatures have been carried out by Bar-Nun et al. (1985b, 1986).

Ultraviolet irradiation and ion bombardment experiments leading to sputtering, heavier molecules, polymers, and dissociation products have been carried out by Greenberg (1982), Brown et al. (1978, 1980, 1982), Lanzerotti et al. (1978, 1982, 1985), Johnson et al. (1982, 1983a, 1983b), Strazulla et al. (1984), de Vries et al. (1984), and Bar-Nun et al. (1985a).

All of the above experiments relate to pure water ice or mixtures of frozen gases. There appear to be no experiments on dust or ice - dust mixtures. Bar-Nun et al. (1985b) find clathrate formation and massive ejection of ice grains when a mixture of amorphous ice is slowly heated. It is not known whether dust grains will be ejected similarly to the clathrates and whether they will be ice covered. It is also not known whether fluffy dust grains will trap frozen gases, whether they will act as a host for some gases more than for others, or how much gas they will trap. This raises many interesting questions. Will a dust mantle over an icy surface (as is apparently the case on the surface of a comet nucleus) trap gases in the dust grains, on the surface of the dust grains, or between dust grains? Can gas escaping through a dust mantle, when cooled, freeze and cause thermal stresses that lead to cracking of the mantle? What are the thermal properties of an ice - dust mixture? How efficient an insulator is a dust mantle that is permeated with frozen gases?

Laboratory measurements of the sublimation of ice and the associated entrainment of dust in insolation experiments will provide important data supplementary to data obtained from planned material release experiments from the space shuttle (Strong et al., 1985). The thermal conductivity of dust - water ice needs to be investigated. Carbonaceous comet dust can be simulated using carbon black particles of various sizes. Various methods to simulate dust particle aggregates and homogeneous mixtures must be investigated. The measurements of thermal conductivity of dust - water ice mixtures need to be compared to clean water ice. These measurements must also investigate several of the parameters involved, such as temperature dependence, dust to ice ratio, dust particle sizes, "fluffy" ice, and physical properties. Additionally, other thermal characteristics must be investigated to see if the dirty water ice parameters vary significantly from clean water ice parameters.

6. COMA MODELS

Considering all models for comets, the coma models are are the most advanced and most successful (see, e.g., Mendis et al., 1985; Huebner et al., 1986). Sublimation (vaporization) of the icy component of a cometary nucleus determines the initial composition of the coma gas as it streams outward and escapes. Photolytic reactions in the inner coma, escape of light species such as atomic and molecular hydrogen, and solar wind interaction in the outer coma alter the chemical composition and the physical nature of the coma gas. Comprehensive models that describe these interactions must include (a) chemical kinetics, (b) coma energy balance, (c) multi-fluid flow for rapidly escaping light components and slower, heavy bulk fluid, (d) separate temperatures for electrons and the rest of the gas, (e) transition from a collision - dominated inner region to free molecular flow in the outer region, (f) dust - gas interaction, (g) mass pick-up by solar wind ions, (h) counter and cross streaming of cometary particles and solar wind ions, and (i) magnetic fields carried with the solar wind.

All of the present coma models assume spherical symmetry. This is a reasonable approximation at large distances in the coma. The following processes are as yet not included in coma models:

The highly asymmetric and irregular dust and gas production that has been observed in comet Halley is not unique to that comet. Although the details of individual jets issuing from the nucleus have only been seen in the close-up pictures taken by the Vega and Giotto spacecraft, the merging of the jets into fans as observed from larger distances and from the ground in comet Halley are very similar to fan-like structures observed from the ground in many other comets. This asymmetric structure must be incorporated in new models of the coma. Pressure equalization in a jet just above the surface of the nucleus will cause a tangential wind across the surface. The number of gas - dust collisions in that direction within the jet is not sufficient to entrain much of the visible dust, although some of the smallest dust grains may be carried with this wind.

A one - temperature representation of the electron energies is not sufficient, as evidenced by the analysis of the data from the flyby of comet Giacobini - Zinner by the International Comet Explorer (ICE) in September 1985. A detailed energy distribution of the electrons or at least a two - temperature representation is needed.

Although ions and electrons are treated as separate fluids from the neutral gas outside the contact surface, ions and neutrals are not treated as separate fluids inside the contact surface. Inward directed flux of ions and electrons has been modeled, but outward directed flux has not been included in comprehensive coma models.

It is not clear whether heterogeneous chemical reactions are important in coma models. It is likely that siliceous dust can act as a catalyst at low temperature and density to form methane and ethane. The importance of this to coma chemistry must be investigated.

Figures 4 to 7 illustrate some of the recent successes of coma models. Figure 4 compares the measurements of electron density, electron temperature, and electron velocity during the flyby of Comet Giacobini - Zinner by the ICE spacecraft. The solid lines are the measured quantities and the dotted lines represent the results from the model calculations of Boice et al. (1986). Figure 5 presents comparisons for the electron density on a larger scale. The experimental data of Bame et al. (1986) has been supplemented with the data from Meyer - Vernet et al. (1986). The smooth solid line is from the comet - solar wind interaction model of Boice et al. (1986) supplemented with calculations from the inner coma model (no solar wind interaction) shown by the dotted extension approaching the Meyer - Vernet measurements asymptotically. Both of these curves are terminated at the point of closest approach of the ICE spacecraft to the comet nucleus. Also shown are the one point calculated by the model of Marconi and Mendis (1986) and two short regions calculated by the model of Fedder et al. (1986). Figure 6 presents a similar comparison on a larger scale for the electron temperature. The observed fluc-

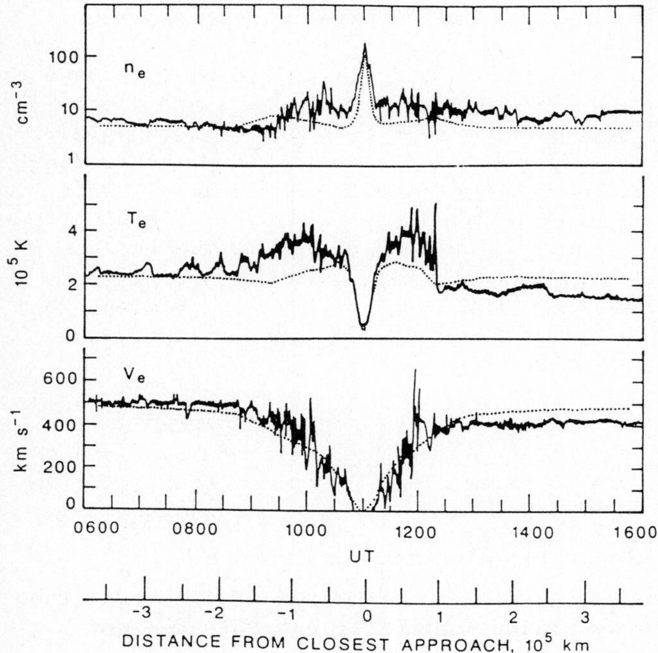

Fig. 4: Density, n_e, temperature, T_e, and flow velocity, V_e, of electrons along the path of intercept of Comet Giacobini - Zinner during flyby of ICE spacecraft.

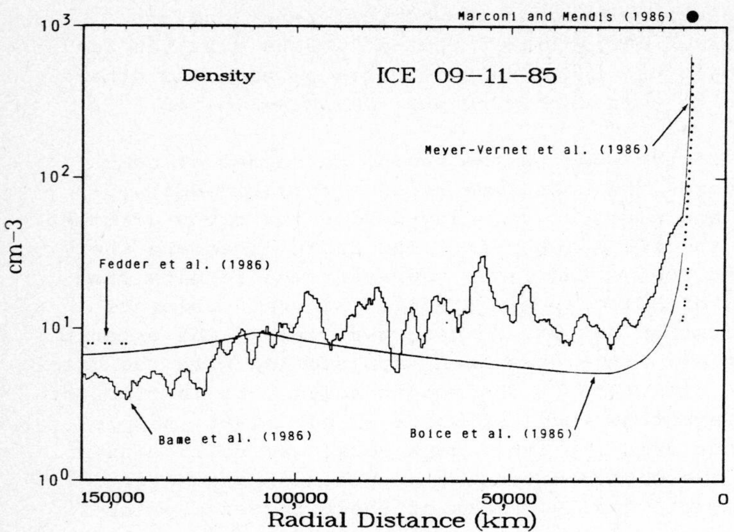

Fig. 5: Measurement and model comparisons of electron density on an enlarged scale along the path of intercept of Comet Giacobini - Zinner during flyby of ICE spacecraft.

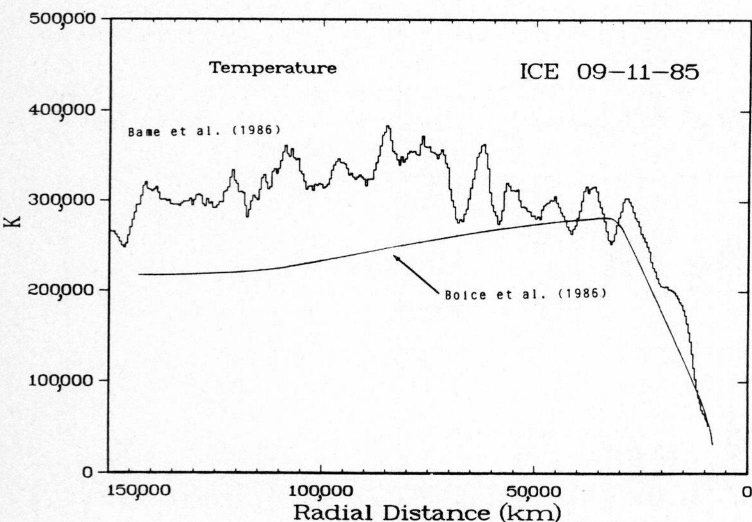

Fig. 6: Same as figure 5, except for electron temperature.

tuations are probably caused by plasma instabilities that have not been included in any model. Finally, in figure 7, the comparison between measured and modeled electron flow velocity is shown on a large scale. One can conclude that these model calculations show good overall agreement, but also have some deficiencies.

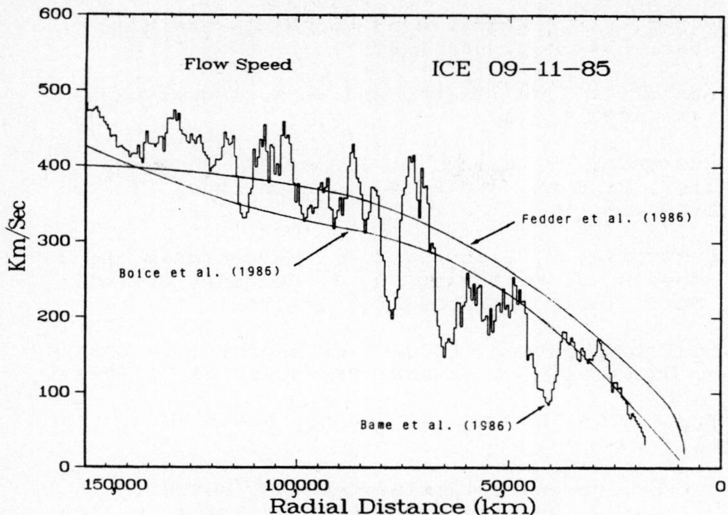

Fig. 7: Same as figure 5, except for electron flow velocity.

7. CONCLUSIONS

Models for the formation of comet nuclei and for the chemico-physical changes in the nuclei while they are in the Oort cloud or the inner comet belt need to be developed. Coma models need to be further improved. The model development should be complemented by laboratory experiments to determine critical parameters and by laboratory and space simulations to provide guidance. Models, laboratory experiments, and laboratory and space simulations are all needed for guidance of instrument development and selection and to support analysis of data obtained from comet missions. Development of models should be part of any mission plan and should be started as soon as possible.

REFERENCES

Bame, S. J., R. C. Anderson, J. R. Asbridge, D. N. Baker, W. C. Feldman, S. A. Fuselier, J. T. Gosling, D. J. McComas, M. F. Thomsen, D. T. Young, and R. D. Zwickl, Science 232, 356 (1986).

Bar-Nun, A., G. Herman, M. L. Rappaport, and Yu. Mekler, Surface Science 150, 143 (1985a).

Bar-Nun, A., G. Herman, D. Laufer, and M. L. Rappaport, Icarus 63, 317 (1985b).

Bar-Nun, A., J. Dror, E. Kochavi, and D. Laufer, Phys. Rev. in press (1986).

Biermann, L. and K. W. Michel, Moon Planets 18, 447 (1978).

Boice, D. C., W. F. Huebner, J. J. Keady, H. U. Schmidt, and R. Wegmann, Geophys. Res. Lett. 13, 381 (1986).

Brown, W. L., L. J. Lanzerotti, J. M. Poate, and W. M. Augustyniak, Phys. Rev. Letters 40, 1027 (1978).

Brown, W. L., W. M. Augustyniak, E. Brody, B. Cooper, L. J. Lanzerotti, A. Ramirez, R. Evat, and R. E. Johnson, Nucl. Instr. Methods 170, 321 (1980).

Brown, W. L., W. M. Augustyniak, E. Simmons, K. J. Marcantonio, L. J. Lanzerotti, R. E. Johnson, J. W. Boring, C. T. Reimann, G. Foti, and V. Pirronello, Nucl. Instr. Methods 198, 1 (1982).

Delsemme, A. H., "Chemical Composition of Cometary Nuclei," in Comets, Ed. L. L. Wilkening, University of Arizona Press, p. 85, (1982).

De Vries, A. E., R. Pedrys, R. A. Haring, A. Haring, and F. W. Saris, Nature 311, 39 (1984).

Donn, B. and J. Rahe, "Structure and Origin of Cometary Nuclei," in Comets, Ed. L. L. Wilkening, University of Arizona Press, p. 203 (1982).

Draganic, I. G., Z. D. Draganic, and S. Vujosevic, Icarus 60, 464 (1984).

Everhard, E., Astrophys. J. 10, L131 (1972).

Fedder, J. A., J. G. Lyon, and J. L. Giuliani, Jr., EOS Trans. AGU 67, 17 (1986).

Festou, C. M., Adv. Space Res. 4, 165 (1984).

Foti, G., L. Calcagno, K. L. Sheng, and G. Strazulla, Nature 310, 126 (1984).

Gail, H.-P. and E. Sedlmayr, Astron. Astrophys. 132, 163 (1984).

Gail, H.-P., R. Keller, and E. Sedlmayr, Astron. Astrophys. 133, 320 (1984).

Greenberg, J. M., "What are Comets Made of? A Model Based on Interstellar Dust," in Comets, Ed. L. L. Wilkening, The University of Arizona Press, p. 131 (1982).

Herman, G. and M. Podolak, Icarus 61, 252 (1985).

Hills, J. G., Astron. J. 87, 906 (1982).

Hills, J. G. and M. T. Sandford II, Astron. J. 88, 1519 (1983a).

Hills, J. G. and M. T. Sandford II, Astron. J. 88, 1522 (1983b).

Huebner, W. F., "The Photochemistry of Comets," in The Photochemistry of Atmospheres, J. S. Levine, Ed., Academic Press, Inc., Orlando San Diego, New York, London, Toronto, Montreal, Sydney, Tokyo, p. 437 (1985).

Huebner, W. F., J. J. Keady, D. C. Boice, H. U. Schmidt, and R. Wegmann, "Chemico-Physical Models of Cometary Atmospheres," in IAU Colloquium No. 120, in press (1986).

Johnson, R. E., L. J. Lanzerotti, and W. L. Brown, Nucl. Instr. Methods 198, 147 (1982).

Johnson, R. E., W. L. Brown, and L. J. Lanzerotti, J. Phys. Chem. 87, 4218 (1983a).

Johnson, R. E., J. W. Boring, C. T. Reimann, L. A. Barton, E. M. Sieveka, J. W. Garrett, K. R. Farmer, W. L. Brown, and L. J. Lanzerotti, Geophys. Res. Letters 10, 892 (1983b).

Klinger, J., Science 209, 271 (1980).

Klinger, J., Icarus 47, 320 (1981).

Klinger, J., "Composition and Structure of the Comet Nucleus and its Evolution on a Periodic Comet," in Ices in the Solar System, Ed. J. Klinger, D. Benest, A. Dollfus, and R. Smoluchowski, D. Reidel Publishing Company, Dordrecht, Lancaster, Boston, p. 81 (1985).

Klinger, J., D. Benest, A. Dollfus, and R. Smoluchowski, Ices in the Solar System, D. Reidel Publishing Company, Dordrecht, Boston, Lancaster (1985).

Kuehrt, E. Icarus, 60, 512 (1984).

Lanzerotti, L. J., W. L. Brown, J. M. Poate, and W. M. Augustyniak, Geophys. Res. Letters 5, 155 (1978).

Lanzerotti, L. J., W. L. Brown, W. M. Augustyniak, R. E. Johnson, and T. P. Armstrong, Astrophys. J. 259, 920 (1982).

Lanzerotti, L. J., W. L. Brown, and R. E. Johnson, "Laboratory Studies of Ion Irradiations of Water, Sulfur Dioxide, and Methane Ices," in Ices in the Solar System, Ed. J. Klinger, D. Benest, A. Dollfus, and R. Smoluchowski, D. Reidel Publishing Company, Dordrecht, Lancaster, Boston, p. 81 (1985).

Marconi, M. L. and D. A. Mendis, Geophys. Res. Lett. 13, 405 (1986).

Mayer, E. and R. Pletzer, "Polymorphism in Vapor Deposited Amorphous Solid Water," in Ices in the Solar System, Ed. J. Klinger, D. Benest, A. Dollfus, and R. Smoluchowski, D. Reidel Publishing Company, Dordrecht, Lancaster, Boston, p. 81 (1985).

Mendis, D. A., H. L. F. Houpis, and M. L. Marconi, Fund. Cosmic Phys. 10, 1 (1985).

Meyer - Vernet, N., P. Couturier, S. Hoang, C. Perche, J.-L. Steinberg, J. Fainberg, and C. Meetre, Science 232, 370 (1986).

Mignard, F. and F. Remy, Icarus 63, 20 (1985).

Patashnick, H., G. Rupprecht, and D. W. Schuerman, Nature 250, 313 (1974).

Pirronello, V., "Molecule Formation in Cometary Environments," in Ices in the Solar System, Ed. J. Klinger, D. Benest, A. Dollfus, and R. Smoluchowski, D. Reidel Publishing Company, Dordrecht, Lancaster, Boston, p. 261 (1985).

Podolak, M. and G. Herman, Icarus 61, 267 (1985).

Prialnik, D., and A. Bar-Nun, "On the Evolution and Activity of Comet Nuclei," preprint (1986).

Remy, F. and F. Mignard, Icarus 63, 1 (1985).

Rickman, H., and C. Froeschle, "Thermal Models for the Nucleus of Comet P/Halley," in International Conference on Cometary Exploration, Ed. T. I. Gombosi, Budapest, Vol. I, p. 75 (1982a).

Rickman, H., and C. Froeschle, "Model Calculations of Nongravitational Effects on Comet P/Halley," in International Conference on Cometary Exploration, Ed. T. I. Gombosi, Budapest, Vol. III, p. 109 (1982b).

Smoluchowski, R., Astrophys. J. Lett. 244, L31 (1981).

Smoluchowski, R., "Amorphous and Porous Ices in Cometary Nuclei," in Ices in the Solar System, Ed. J. Klinger, D. Benest, A. Dollfus, and R. Smoluchowski, D. Reidel Publishing Company, Dordrecht, Lancaster, Boston, p. 81 (1985).

Strazulla, G., V. Pirronello, and G. Foti, Astron. Astrophys. 123, 93 (1983).

Strazulla, G., L. Calcagno, and G. Foti, Astron. Astrophys. 140, 441 (1984).

Strong, I. B., R. R. Brownlee, E. H. Farnum, W. F. Huebner, T. D. Kunkle, J. R. Stephens, and M. F. Bode "Plans for Release of Simulated Interplanetary Materials into Low Earth Orbit," in IAU Colloquium No. 85, (1985).

Tscharnuter, W. M., Space Sci. Rev. 27, 235 (1980).

Wallis, M. K., Nature 284, 431 (1980).

Weissman, R. P., and H. H. Kieffer, Icarus 47, 302 (1981).

Weissman, R. P., and H. H. Kieffer, J. Geophys. Res. LPI, C358 (1983).

Spatial Distribution of Constituents in the Coma of Comet Halley, an Observing Programme at the ESO 1-m Telescope

K. Jockers[1], *E.H. Geyer*[2], *and A. Hänel*[2]

[1]Max-Planck-Institut für Aeronomie,
D-3411 Katlenburg-Lindau, Fed. Rep. of Germany
[2]Observatorium Hoher List, D-5568 Daun, Fed. Rep. of Germany

1. Scientific Objective

When a comet visits the inner solar system the sun's radiation sublimates part of the nucleus' matter. The liberated gases drag some dust particles with them and form the coma of the comet. It has an approximately spherical shape and an extent of several hundred thousand kilometres. The sublimated gas particles are dissociated and ionized by solar UV radiation, charge exchange and, in the inner coma, by collisions. In the inner coma chemical reactions between the different coma species will form new types of radicals and ions. Ultimately, all gas molecules and their daughter products will be ionized and swept away by the solar wind into the cometary ion tail. The dust particles are removed from the coma by solar radiation pressure and form the dust tail.

The observing programme to be described in the following was devoted to a study of the different constituents in the cometary coma. Such a study should give information on the chemical composition of the cometary nucleus. The neutral radicals, which are observable from the ground like CN, C_2, C_3, CH, NH and NH_2, are chemically processed and therefore relate only indirectly to the so-called mother substances of the nucleus. Many ions, however, e.g. CO^+, CO_2^+ and H_2O^+, are ions of presumable nucleus constituents. Consequently, the interest concentrated on the cometary ions. Their behaviour, however, is influenced by their interaction with the solar wind, which leads to the formation of ion rays and streamers. Therefore, a study of cometary ions must include their kinematical behaviour. As Comet Halley was investigated by an armada of space probes, we have the unique opportunity to compare the ground-based observations with in situ measurements.

2. The Instrument

For the observations the focal reducer of the Observatory Hoher List was used at the ESO 1-m telescope. For the comet observations the

Max-Planck-Institute for Aeronomy supplemented this instrument with two dioptric cameras (Carl Zeiss, Oberkochen) for the near UV (365 - 500 nm) and visual (425-660 nm) spectral ranges. A two-stage proximity-focused image intensifier (Proxitronic, Bensheim) with bialkali cathode was attached to the UV camera and the image was recorded on plates (mostly hypersensitized IIIa-F) pressed against the exit window of the intensifier. The optical arrangement is shown in Figure 1. The telescope beam behind the Cassegrain focus (a) is recollimated via a field lens (b) and a collimator triplet (c), and a new image, reduced in size by a factor of 5, is formed by the lens (g) on the photocathode of the image intensifier (h). At the 1-m telescope a field of 25 arcminutes is obtained which corresponds at Comet Halley to about 10^6 km (depending on its geocentric distance).

The instrument was used in three modes. In the imaging mode (Figure 1) pictures of the comet were obtained through interference filters combined with a tunable narrow-gap Fabry-Perot filter (Queensgate Instruments, Sunbury near London, B. Halle, Berlin). The Fabry-Perot works in the wavelength interval of 350-430 nm. It has a bandpass (FWHM) of about 6 Å and a free spectral range of about 100 Å. In the field spectroscopy mode (Geyer et al., 1979) a slit mask with a pattern of 70 0.2 mm wide slits was inserted into the Cassegrain focal plane. Instead of the Fabry-Perot filter a direct vision grating prism was put into the parallel beam to produce 70 simultaneous spectra at different places in the cometary coma. The resulting saving of observing time was essential for the success of the programme, in particular when the comet was still close to the sun. Two gratings with 300 and 600 lines/mm were used and gave an inverse dispersion of 207 and 103 Å/mm, respectively. One plate was obtained with a double grating prism. The resulting pairs of inverted spectra should allow the determination of radial velocities (Geyer and Nelles, 1985). In a third

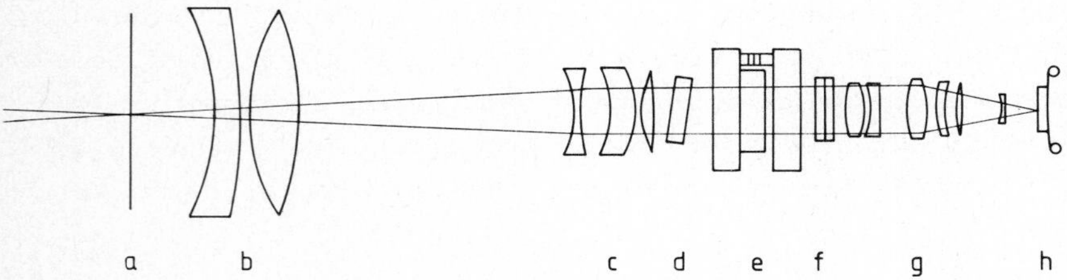

Figure 1: Optical arrangement of the focal reducer in the imaging mode. a: Cassegrain focus; b: field lens; c: collimator lens; d: coloured glas prefilter; e: tunable Fabry-Perot filter; f: interference filter; g: camera lens f/2.8; h: two-stage image intensifier.

mode, another Fabry-Perot etalon with a fixed plate separation of 0.5 mm was added to the optical arrangement of Figure 1 in an attempt to derive Doppler velocities of the cometary ions. A few very weakly exposed interferograms were obtained in the light of the CO^+ and CH^+ ions. It seems questionable if they will allow derivation of ion speeds. Most plates exposed in the direct imaging and field spectroscopy modes were photometrically calibrated with the ESO spot sensitometer, with exposures of the bipolar nebula NGC 6302 and with a set of mercury standard lamps. Besides the focal reducer, three cameras, attached to the top ring of the telescope, were used to obtain wide-field images and slitless wide-field spectra in the visible and UV spectral ranges.

3. The Observations

The observations were performed in the two periods March 10-16 and April 4-11, 1986. As there has not been enough time yet for a detailed quantitative analysis of the data, only some raw data are presented. Figure 2 shows a pair of images taken on March 15 at 367.4 nm (CO_2^+) and 365.0 nm ("continuum"). Both images were taken through the same interference prefilter but correspond to different settings of the Fabry-Perot filter. Comparison of the two images indicates a strong signal at the wavelength of CO_2^+. The weak ion features present in the "continuum" picture may be caused by weak plasma emissions in the "continuum" window or, more likely, by spectral impurity of the

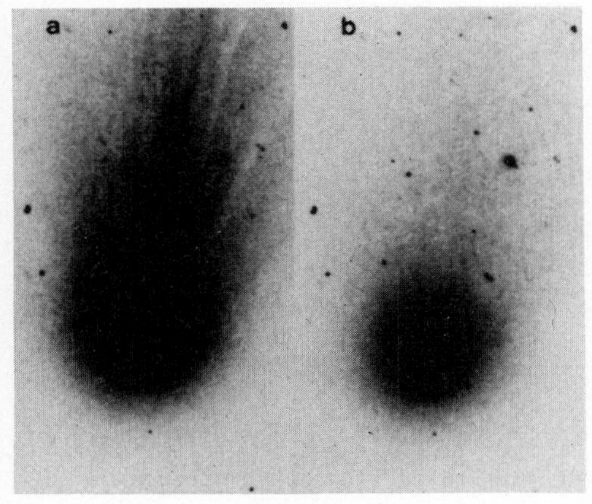

10 arcmin

4.1×10^5 km

Figure 2: Images of Comet Halley obtained March 15. a: CO^+ at 367.4 nm; b: "continuum" at 365.0 nm. Exposure times: 10 minutes.

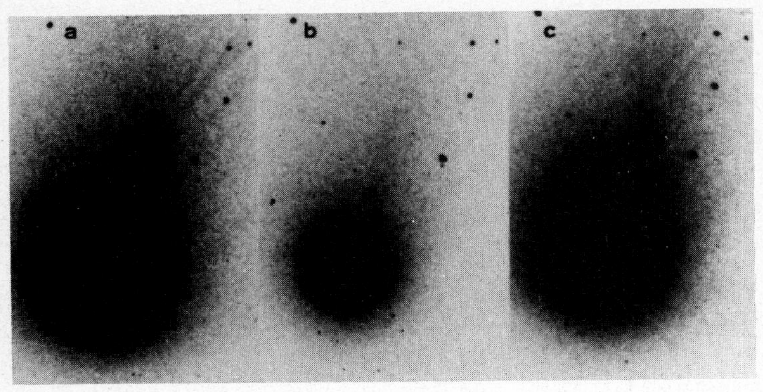

10 arcmin

3.9×10^5 km

Figure 3: Images of Comet Halley obtained March 16. a: CO^+ at 401.9 nm; b: "continuum" at 407.4 nm; c: N_2^+ at 391.2 nm. Exposure times: a and b: 1 minute; c: 4 minutes.

Fabry-Perot filter. Note the regular pattern of ion streamers surrounded by a plasma envelope which is missing in the continuum picture. In Figure 3 we see three images, obtained at 401.9, 407.4 and 391.2 nm respectively, corresponding to the 3-0 $A^2\,\pi_{3/2} - X^2\Sigma^+$ transition of CO^+, "continuum" and the 0-0 $B^2\Sigma^+u-X^2\Sigma^+g$ transition of N_2^+. The N_2^+ emission is present but weaker than CO^+ and N_2^+. There is a neutral gas coma visible on the CO^+ and N_2^+ pictures which is due to C_3 and CH, respectively. In the second observation period the comet was observable almost all night. The development of the inner tail of the comet during 6 hours is illustrated in Figure 4. Dramatic changes are seen in the light of the CO^+ and CO_2^+ ions. The images in the light of the two molecular ions look similar but there seem to be systematic differences. They need not to correspond to differences in column density ratio but may be caused by Greenstein effect (change in fluorescence efficiency caused by Doppler shift of the solar spectrum as seen by the moving cometary ions). An example of a multi slit spectrum is presented in Figure 5. It covers the region between 350 and 430 nm. To show the location of the comet relative to the slits, with most spectra a calibration exposure was obtained by removing the grating and taking a double exposure. Direct images of the slits were exposed, then the slit mask was removed and an image of the comet was taken through an interference filter of 10 nm passband centred at 369 nm. The spectra show all major ions except H_2O^+ which has no emissions between 350 and 430 nm. Note the varying line ratios in the different spectra. The ion emissions are also seen in spectra which do

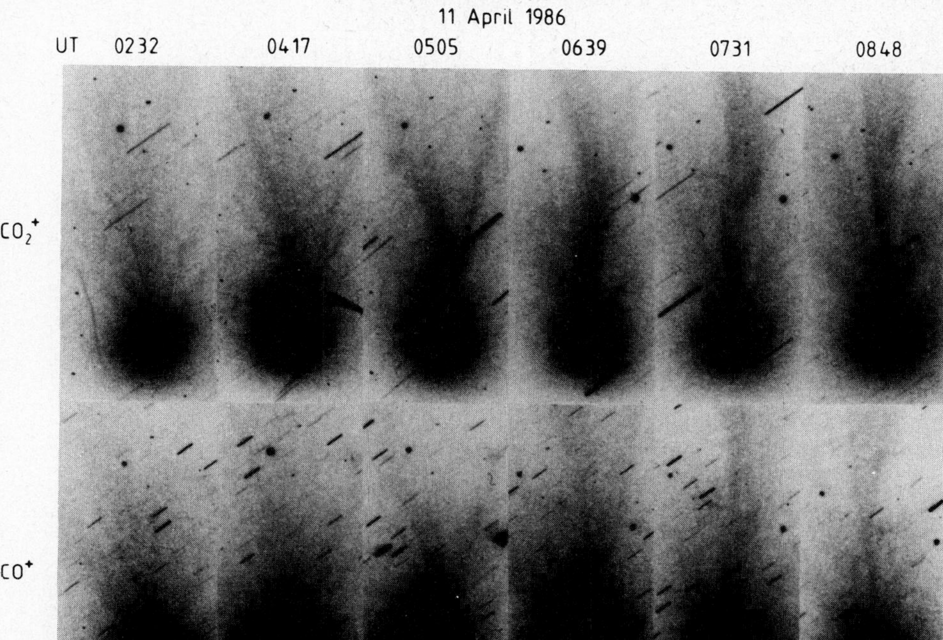

Figure 4: Motions in the inner plasma tail of comet Halley recorded in the light of the CO_2^+ and CO^+ ions (April 11). Exposure times: CO_2^+: 13 minutes; CO^+: 4 minutes.

not correspond to a visible ion streamer. CO^+ ions are even seen upstream of the comet, confirming the notion of an extended CO^+ ion source region.

Figure 6 shows an example of a wide-field slitless tail spectrum taken with the Zeiss UV Sonnar with 104 mm focal length. The spectrum was taken through a UG 5 filter and covers the wavelength range from 309 to 395 nm. A direct, unfiltered, blue image of the comet is presented at the same scale for comparison. The wide-field images were taken simultaneously with the two latest pictures of Figure 4 but the spatial scale is almost 10 times larger. The extended CN coma in the wide-field picture would cover almost half of the frames of Figure 4.

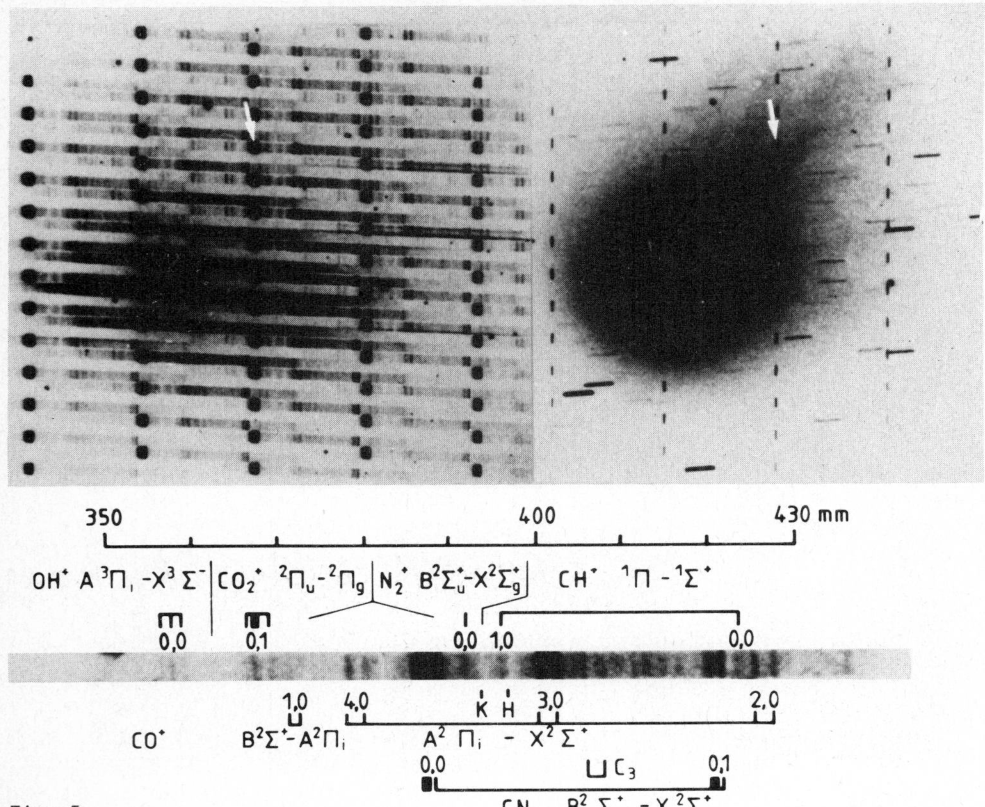

Fig. 5

Fig. 6

Figure captions see opposite page.

Acknowledgements

We are indebted to W.I. Axford, H. Rosenbauer and V.M. Vasyliunas for a generous support of the ground-based comet camera project. The work would not have been possible without the helping hands of engineers and workers in the shops of the Max-Planck-Institute of Aeronomy and the Observatory Hoher List. We are very grateful for the possibility to conduct the observations under the cloudless skies of the European Southern Observatory and for the invaluable assistance provided by ESO staff at La Silla and Garching.

K. Jockers would like to thank Dr. H.U. Schmidt on occasion of his 60th birthday for having him introduced to the field of cometary plasma tails.

References

Geyer, E.H., Hoffmann, M., Nelles, B.: 1979, Astron. Astrophys. 80, 248

Geyer, E.H., Nelles, B.: 1985, Astron. Astrophys. 148, 312

Lund, G., Surdej, J.: 1986, Messenger 43, 1

This article also appeared in The ESO Messenger No 44, 1986. The editors thank ESO for providing the photographs.

Figure 5. Multi-slit spectrum of Comet Halley obtained on April 10. Top left: Spectrum plate; top right: slit position plate. The arrows point to the slit and the corresponding spectrum, which is reproduced in enlarged form in the bottom to explain the individual spectrum features. Exposure time of spectrum: 30 minutes.

Figure 6: Slitless wide-field spectrum and direct photograph of Comet Halley. Wavelength range of spectrum 309-395 nm. Dispersion E-W. Wavelength increases to the right. Exposure time of spectrum: 1 hour at f/4.5 on IIa-0 plate.

On the Average Chemical Composition of Cometary Dust

E.K. Jessberger[1], *J. Kissel*[1], *H. Fechtig*[1], *and F.R. Krueger*[2]

[1]Max-Planck-Institut für Kernphysik, P.O. Box 103980, D-6900 Heidelberg, Fed. Rep. of Germany

[2]MPI-Consultant, D-6100 Darmstadt, Fed. Rep. of Germany

This is a progress report on our efforts to extract information from mass spectra obtained by the particulate impact analysers onboard the VEGA and GIOTTO spacecrafts. Analysing a subset of 23 selected spectra obtained with identical instrument conditions, we derive a preliminary average abundance pattern of the major and some minor elements in Halley's dust. Within a factor of two the pattern is chondritic for the major elements. Carbon is enriched by a factor of 8 compared to Cl chondrites. Halley's dust is composed of chondritic "silicates" and refractory carbonaceous material.

Keywords: cometary dust, elemental abundance, chondrite, PIA, PUMA, organics, density.

INTRODUCTION

The two VEGA and the GIOTTO space probes for the first time allowed the elemental and isotopic analysis of cometary dust. On each of the three missions a mass spectrometric particle analyser was flown: PIA onboard GIOTTO and PUMA 1,2 onboard VEGA 1,2 (Refs. 1,2). The experiments will briefly be described below. First glance results have been published before (Refs. 3,4).

For the proper design of an analyser for such almost totally unknown dust particles some concept on the dust properties was needed. Grain size distribution and dust fluxes were modelled after optical data (Ref. 5). With regard to the dust composition, it was commonly anticipated knowledge that comets are "primitive" bodies. The most primitive material on earth are Cl carbonaceous chondrites and chondritic interplanetary dust particles, IDPs (Refs. 6,7). So it seemed natural to expect similar material in a comet. IDPs, and especially the low-density fluffy ones (Fig.1), are regarded at least as analogs to cometary dust if they were not of cometary origin for which recently evidence is growing (Ref. 8). Probably, however, they are not pristine cometary particles but altered either in space (Ref. 9) or during entry in the Earth's atmosphere (Refs. 10,11).

Anders (Ref. 12) has stressed that Cl chondrites most probably are not directly linked to comets but that "meteorites provide a good concep-

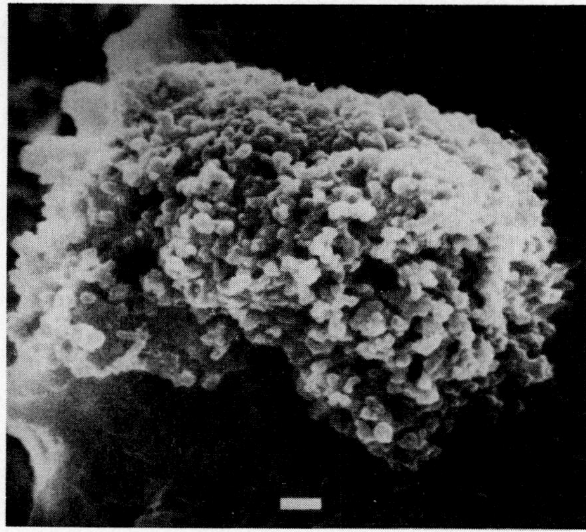

Figure 1: Interplanetary dust particle collected in the stratosphere.

tual framework for understanding comets" (Ref. 13). Table 1 lists the abundances of some elements in Cl and the slightly more evolved C2 chondrites (Ref. 14-16), the primordial cosmic abundance (Refs. 17,18 and a proposed composition of a comet (gas and dust) (Ref. 19). Differences between these compositions concern mostly the light volatile elements CHON where the solid matter (Cl, C2) is depleted. The depletion is correlated to the degree of planetary reprocessing and therefore should be small in supposedly primitive comets. In addition to minerals, Cl chondrites and IDPs (Ref. 8) contain carbonaceous components which were also expected to be present in cometary dust.

THE EXPERIMENT

The particulate impact analysers have been described in detail before (Refs. 1,2) and here only a short overview is given. Fig.2 shows a schematic of the experiment. Dust grains impact on a movable silver or silver-doped platinum target with a speed of 69 km/sec (GIOTTO) and 78/79 km/sec (VEGA 1,2), respectively. Upon impact, the grain and some of the target material is vapourized and partly ionized. Positive ions are accelerated with 1000 V into a field-free drift tube. Ions with small m/e-ratio travel with higher speed than those with larger m/e resulting in a time-of-flight mass spectrum at the detector at the end of the drift tube. This tube consists of two sections with an electrostatic ion reflector in between. The reflector increases the mass resolution to $m/\Delta m \sim 150$.

Development and tests of the instrument have been performed using the Heidelberg Dust Accelerator. Due to the nature of this facility (see e.g. Ref. 20), only a limited variety of projectile materials could be

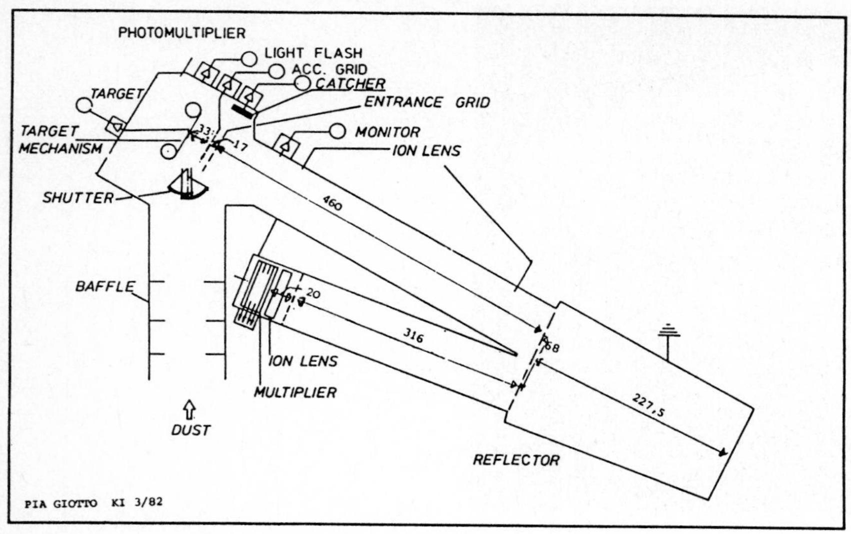

Figure 2: Schematic of the particulate impact analyzers onboard the VEGA 1,2 and GIOTTO spaceprobes.

used. Speeds up to 64 km/s have been obtained, but the majority of measurements was done at speeds of 15-50 km/s. From these studies it became apparent (Ref. 21) that the amount of target- and projectile ions depend on the particle's mass, and its density: The number of silver ions (from the target) is a measure of projectile mass; the ratio of projectile ions to target ions is a measure of projectile density (mass per unit volume of envelope) where an increasing ratio is associated with decreasing projectile density.

The major problem of the experiment was the derivation of ionization yields for all relevant elements. As has been stated above, there is no accelerator to calibrate the instrument, i.e. to accelerate complex projectiles to $\sim$70-80 km/sec. From the experience with < 50 km/sec particles it was assumed that at 70-80 km/sec most projectile molecules are destroyed and that mostly atomic ions appear in the mass spectra. It was shown (for speeds < 50 km/sec) that the relative ion yields for dust particle impact are well comparable with those obtained with other techniques using fast deposition of energy on solid surfaces, e.g. SIMS, LID, EPID (Ref. 22). Consequently, for the present analysis these ion yields are applied to quantitatively build up synthetic mass spectra for comparison with the cometary mass spectra.

RESULTS

For the present communication we analyzed 23 spectra from PUMA 1. These are <u>all</u> the spectra transmitted in a transient-recorder-like mode under identical operation conditions of the instrument (high sensitivity, low reflector voltage) (for details see Refs. 1,2). Thus, the selection is unbiased. The spectra were taken over the whole time period of 25 minutes around VEGA1 encounter.

By extrapolation from laboratory experiments (Refs.1,21) we derive densities in the range 0.4 g/cm^3 to 0.01 g/cm^3 (mean 0.1 g/cm^3) which is close to values derived from meteor phenomena (Refs. 23,24) and indeed reflects the fluffy nature of cometary particles (cf. Fig.1). The particle diameters are estimated to vary from 0.02 µm to 5 µm. Five out of the 23 particles appear to be larger than 1 µm. The total mass analyzed in the spectra is about 4 x 10^{-11}g.

Fig. 3 shows the main sections of two typical spectra together with the simulated spectra (see above). They demonstrate that most of the

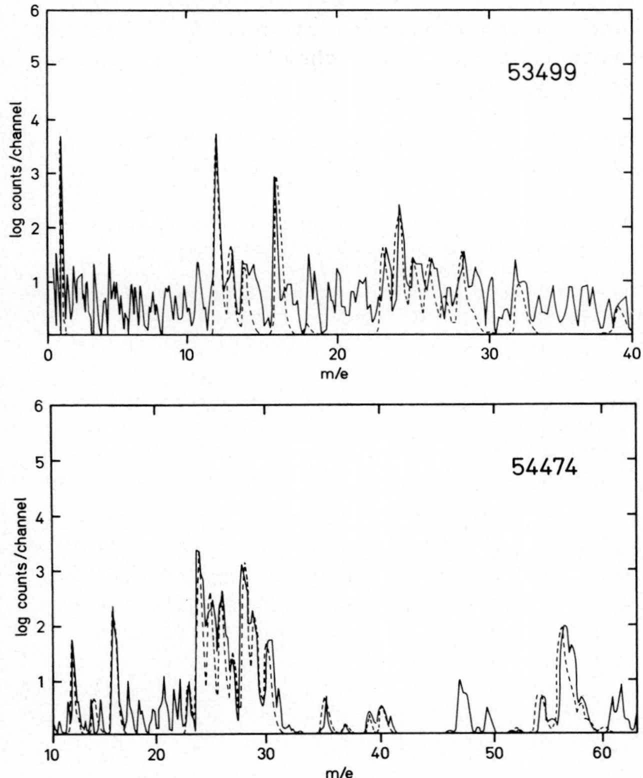

Figure 3: Informative portions of time-of-flight spectra obtained by PUMA 1 onboard VEGA 1. The solid lines are the measured spectra, the dotted lines the fit spectra. The fit involves only atomic ions and uses normal isotopic composition of the elements.
53499: Above m/e=40 only the Ag peaks (80 cts/ch) are used. Estimated density 0.01 g/cm^3 and diameter < 0.1 µm.
54474: Below m/e=10 no peak > 5 cts/ch appears and above m/e=63 only the Ag peaks (600 cts/ch). Estimated density 0.1 g/cm^3 and diameter 0.4 µm.

peaks can be regarded as atomic ions taking the normal isotopic composition of the elements, e.g. C in No. 53499 and Mg and Si in No. 54474, where even the low intensity peaks of Cℓ fit very nicely. Some peaks, however, are difficult to explain and their similar peak heights point to a common origin, possibly background from an unknown source. But it is remarkable that the preflight contention that atomic ions would dominate the PIA/PUMA mass spectra (Ref. 21) is confirmed by the data.

The two spectra differ markedly in the ratio of the light volatile elements (CHON) to the heavy elements (MgSiFe). In terms of ion counts this ratio is 24 for No. 53499 and for No. 54474 it is 0.07. Thus, these spectra examplify the extremes of particle types we encountered, one enriched in volatiles and one where these elements are depleted (Refs. 3,4). In 10 of the 23 spectra the count ratio C/Si is larger than 0.5 (range 0.5 to 113) and in the remaining spectra the ratio is less than 0.2. This demonstrates that the gross chemical composition of cometary dust particles is very non-uniform.

Despite these large variations, the subset of spectra can be viewed to represent the average chemical composition of Halley's dust. To deduce

Table 1: Abundances of the elements in the primordial solar system (cosmic abundances) and in carbonaceous chondrites type 1 and 2. The stated uncertainty (1 σ) reflects the variability within each group. The proposed cometary abundances are for dust and gas. The last column gives the abundance pattern of Halley's dust relative to that of C1 chondrites which are also the sole source of the stated uncertainties (1 σ).

	primordial solar		C1 chondrites errors reflect the variability	C2	Ref.	proposed cometary (Ref. 19)	PUMA I norm. to Si and Cℓ (this work)
	Ref. 17	Ref. 18					
H	2.5×10^6	2.7×10^6	590 ± 50	308 ± 56	16	1500	.025 ± .002
C	790	1210	76 ± 13	38 ± 2	16	300	7.9 ± 1.4
N	210	248	6.1 ± .2	2.3 ± .4	16	> 10	3.4 ± .1
O	1700	2010	730 ± 150	530 ± 50	14	2100	.44 ± .09
Na	5.7	5.7	12 ± 9	4 ± 1	15		2.2 ± 1.7
Mg	101	108	103 ± 7	104 ± 5	15		.48 ± .03
Al	8.0	8.5	10.3 ± 3.2	9.4 ± .6	15		.46 ± .14
Si	≡100	≡100	≡ 100	≡ 100		≡ 100	≡ 1
S	48	51.5	48 ± 10	22 ± 7	15		1.6 ± .3
Cℓ	.47	.52	.32 ± 17	.21 ± .1	14		> 3000
K	.35	.38	.7 ± .6	.3 ± .1	15		.19 ± .16
Ca	5.9	6.1	7.1 ± 1.2	7.1 ± 1.0	15		.40 ± .07
Ti	.24	.24	.29 ± .06	.24 ± .06	15		3.4 ± .7
Cr	1.4	1.34	1.0 ± .3	1.2 ± .1	15		1.2 ± .4
Mn	.87	.95	.87 ± .15	.6 ± .1	15		1.6 ± .3
Fe	86	90	86 ± 4	83 ± 3	15		.77 ± .04
Ni	4.8	4.93	4.4 ± .7	4.6 ± .3	15		.53 ± .08

the elemental abundance, we sum the counts for each element and calculate with the relative ionisation yield given by (Refs. 21,22,25) the atomic abundance. The ion yields are considered to be accurate within at least a factor of two with the possible exceptions of H and Cℓ which may bear larger uncertainties. It should be pointed out that molecular ions or doubly charged ions are not considered with the consequence that the calculated abundances are upper limits. The Si-normalized abundances, relative to Cl-chondrites, are given in Table 1.

The stated uncertainties are derived exclusively from the variation in Cl composition (Ref. 15) and not from PUMA data.

Hydrogen appears to be greatly depleted in the dust or else the ion yield used is not applicable. N and Ti seem to be overabundant but it has to be taken into account that for both, because of the low ion yield, the total count number is 100 times less than that of Mg and is the smallest of all considered elements. Na, S, Cr, and Mn are indistinguishable from Cl within 2 sigma. The major elements O, Mg, Al, Ca, and Fe as well as Ni seem to be depleted by an average value of 0.55 $\pm$ .13 relative to Cl and Si. This constant factor may suggest that the estimate of the Si-ion yield is wrong by a factor of two. Up to now, however, there is no other supporting evidence and the question remains open. In any case, it is remarkable, though expected by some, that the major rock-forming elements within a factor of two have the same abundance pattern as the Cl chondrites and the sun. Carbon is enriched relative to Cl by a factor of 8 but is very similar to the cosmic abundance. This relatively high C-abundance points to the presence of a refractory carbonaceous component, since ices can be thermodynamically excluded at the Halley-VEGA distance (Ref. 26).

The dust of comet Halley is composed of silicates with cosmic composition and carbonaceous material and appears to be similar to chondritic interplanetary dust particles (Ref. 27). The present status of data evaluation does not yet allow to decide on elaborate experimental dust models (Refs. 28,29), nor on the presence or on special formation processes of organic material (Ref. 29-31) and therefore not on conceptions of larger consequences (Ref. 32).

REFERENCES

1. Sagdeev R Z et al. 1985, Venus-Halley Mission, Louis-Jean, Gap

2. Kissel J 1986, Eur Space Ag Spec Publ 1077, 67-68.

3. Kissel J et al. 1986, Composition of comet Halley dust particles from Vega observation. Nature 321, 280-282.

4. Kissel J et al. 1986, Composition of Halley dust particles from Giotto observations. Nature 321, 336-338.

5. Divine N 1981, Eur Space Ag Spec Publ 174, 25-30.

6. Anders E 1971, How well do we know "cosmic" abundances? Geochim Cosmochim Acta 35, 516-522.

7. Brownlee D E 1985, Cosmic dust: collection and research. Ann Rev Earth Planet Sci 13, 147-173.

8. Bradley J P & Brownlee D E 1986, Cometary particle: thin sectioning and electron beam analysis. Science 231, 1542-1544.

9. Mukai T & Fechtig H 1983, Packing effect of fluffy particles. Planet Space Sci 31, 655-658.

10. Fraundorf P 1980, Microcharacterization of interplanetary dust collected in the Earth's stratosphere. Ph D thesis, Washington Univ St. Louis, USA.

11. Jessberger E K & Wallenwein R 1986, PIXE characterization of stratospheric micrometeorites. Adv Space Res (in press).

12. Anders E 1975, Do stony meteorites come from comets? Icarus 24, 363-371.

13. Anders E 1986, What can meteorites tell us about comets? To appear in: "Comet Nucleus Sample Return", Proc ESA Workshop, Canterbury, U.K.

14. Mason B 1971, Handbook of Elemental Abundances in Meteorites. Gordon and Breach Science Publ NY, 555 pages.

15. McSween H Y Jr & Richardson S M 1977, The compostions of carbonaceous chondrite matrix. Geochim Cosmochim Acta 41, 1145-1161.

16. Van Schmus W R & Hayes J M 1974, Chemical and petrographic correlations among carbonaceous chondrites. Geochim Cosmochim Acta 38, 47-64.

17. Palme H, Suess H E & Zeh H D 1981, Abundances of the elements in the solar system. In: Landolt-Börnstein 2a (eds K Schaifers, H H Voigt), pp 257-272.

18. Anders E & Ebihara M 1982, Solar system abundances of the elements. Geochim Cosmochim Acta 46, 2363-2380.

19. Delsemme A H 1977, The pristine nature of comets. In: "Comets, Asteroids, Meteorites" (ed. A H Delsemme, U. of Toledo Press), pp 3-13.

20. Fechtig H, Grün E & Kissel J 1978, Laboratory simulation. In: "Cosmic Dust" (ed. J A M McDonnell), Wiley, Chichester, pp 607-667.

21. Krueger F & Kissel J 1984, Experimental investigations on ion emission with dust impact on solid surfaces. ESA SP-224, 43.

22. Kissel J & Krueger F 1986, Ion formation by impact of fast dust particles and comparison with related techniques. Appl Phys A (in press).

23. Ceplecha Z 1977, Meteoroid populations and orbits. In: "Comets, Asteroids, Meteorites" (ed. A H Delsemme), U of Toledo Press, pp 143-152.

24. Verniani F 1969, Structure and fragmentation of meteoroids. Space Sci Rev 10, 230-261.

25. Krueger F R 1986, Ion emission from insulators during impact of accelerated dust particles and comparison with other techniques. Radiation Effects (in press).

26. Hanner M S 1981, On the detectability of icy grains in the comae of comets. Icarus 47, 342.

27. Fraundorf P, Brownlee D E & Walker R M 1982, Laboratory studies of interplanetary dust. In: "Comets" (ed. L L Wilkening) U of Arizona Press, pp 383-409.

28. Greenberg J M 1982, What are comets made of? A model based on interstellar dust. In: "Comets" (ed. L L Wilkening) U of Arizona Press, pp 131-163.

29. Greenberg J M 1983, Interstellar dust, comets, comet dust and carbonaceous meteorites. In: "Asteroids, Comets, Meteors" (eds. C I Lagerkvist, H Rickman) pp 259-268, Uppsala Univ Press.

30. Strazulla G 1985, Modifications of grains by particle bombardment in the early solar system. Icarus 61, 48-56.

31. Rössler K, Jung H-J & Nebeling B 1984, Hot atoms in cosmic chemistry. Adv Space Res 4, 83-95.

32. Hoyle F & Wickramasinghe Ch 1981, Comets - a vehicle for panspermia. In: "Comets and the Origin of Life" (ed. C Ponnamperuma), D Reidel Publ Comp, pp 227-239.

The Interaction of the Solar Wind with a Comet

R. Wegmann

Max-Planck-Institut für Physik und Astrophysik, Institut für Astrophysik, Karl-Schwarzschild-Str. 1, D-8046 Garching, Fed. Rep. of Germany

In 1951 Biermann [2] detected the solar wind as the agent, which shapes the cometary plasma to an extended tail. Numerical model calculations for the interaction of the solar wind with a comet started in 1967 with a simplified one-dimensional model [3]. Since then the model prospered and matured from hydrodynamics to MHD and grew from one to three dimensions [5]. A comprehensive presentation of the results was given by Schmidt and Wegmann [6]. Recently the model got new impact from the collaboration with the group of Huebner in Los Alamos [4]. Last year it had to face the major challenge posed by the ICE spacecraft, which made the first in situ measurements of a cometary plasma. I report here on some improvements which we included recently in this "Schmidt-Wegmann-model". Furthermore I present some results, which we obtain for a comet like comet Giacobini-Zinner and compare these with the data from the ICE spacecraft.

Before we can study the interaction, we must consider the question of how a comet would look like if there were no solar wind. The nucleus of a comet is a solid body of several kilometers in diameter consisting of ice and dust. Heated by the solar radiation this nucleus releases several tons of gas and dust per second. The gas consists of originally neutral atoms and molecules. These get a velocity of about 1 km/sec and are in the course of time transformed by various chemical and ionisation processes. In the absence of solar wind the ions move together with the neutrals.

Huebner, Keady and Boice have made model calculations for this situation. From certain assumptions about the size of the nucleus (radius = 1 km), its albedo and its chemical composition they conclude that comet Giacobini-Zinner releases $4.5 \cdot 10^{28}$ particles per second. Figure 1 shows the number densities of a few selected species in dependence of the distance from the nucleus. The neutral gas consists mainly of water. There are also rather complex molecules. I have

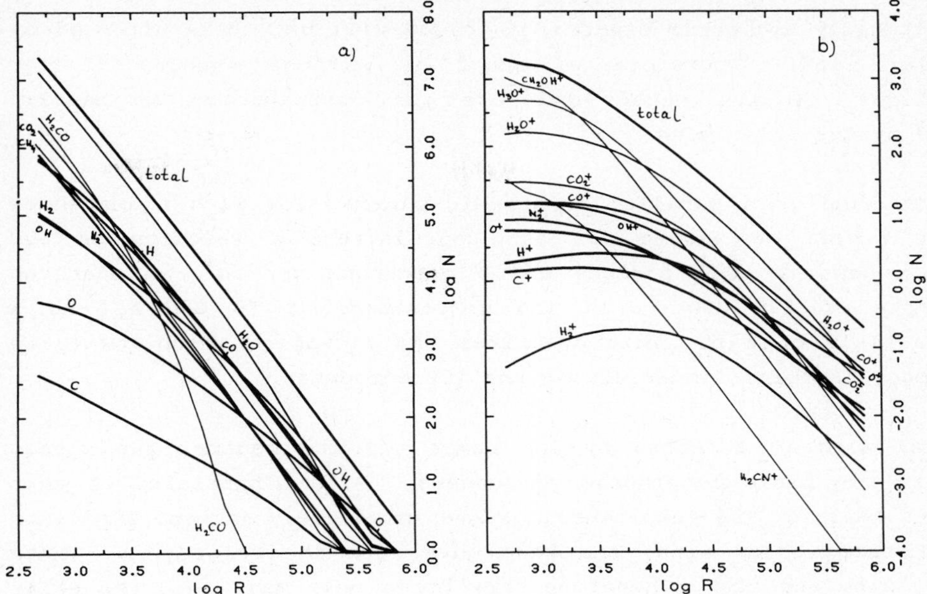

Fig.1. Number densities for a spherically symmetric model in dependence of the distance R from the nucleus. a) neutrals, b) ions. (Courtesy, W.F. Huebner et al.)

distinguished in the figures the curves pertaining to atoms, diatomic, triatomic and polyatomic molecules by the line thickness. It becomes apparent that the densities decrease for complex molecules more rapidly with distance from the nucleus. This is caused by the fact that the molecules are less stable than the atoms. The trend is from complexity to simplicity.

For all these neutrals there are also the corresponding ions (fig. 1b). In the outer part there are mainly the ions of water, oxygen, carbon monoxide and carbon dioxide. In the dense inner parts there are also ions like H_3O^+, which are generated by chemical processes. In the inner part the neutral density is by several orders of magnitude higher than the ion density (The scales for the ordinates in fig.1 are different!).

The density of the neutrals decreases like R^{-2} with distance R from the nucleus. As this gas gets ionized, it provides a supply for the ions. Therefore, the ion density decreases more slowly with distance. In the inner parts the behavior is like R^{-1}. The time-scale for ionisation is about $3 \cdot 10^6$ seconds. Therefore only outside a sphere of several million kilometers the density of the ions becomes comparable to that of the neutrals.

The spherically symmetric neutral gas cloud will be little affected by the solar wind. Therefore we adopt it without changes in our calculations. On the other hand the ion distribution is heavily affected by the solar wind.

The solar wind is a parallel supersonic plasma flow with an embedded magnetic field. We assume for our calculations a velocity of 500 km/sec, a density of 5 protons and 5 electrons per cm^3, an electron temperature of 230 000 K, an ion temperature of 50 000 K, and a magnetic field of 8 nT. This describes the situation which prevailed near comet Giacobini-Zinner during the ICE encounter.

The solar wind is affected by the comet via its neutral gas cloud. There are two basic interaction processes. First, particles of this cloud are ionized. The resulting ions are immediately incorporated into the magnetized solar wind, and then move with the solar wind. These ions are slow and cool. Therefore they bring only mass into the solar wind, but neither momentum nor energy.

Secondly, there are elastic collisions of the solar wind ions and the neutrals. This process contributes to the momentum and energy balance of the plasma. Let N_n be the number density of the neutrals, v their velocity, and m_n the mean mass of a neutral particle. Correspondingly, let N_i, u and m_i be the number density, velocity, and mean mass of the ions. Then the neutrals exert by elastic collisions a volume force

$$\dot{I} = F \frac{m_i m_n}{m_i + m_n} (\vec{v} - \vec{u}) \tag{1}$$

on the ions. The frequency of collisions

$$F = Q \; N_n \; N_i \sqrt{|u-v|^2 + \gamma p/\rho} \tag{2}$$

is proportional to the relative velocity. The thermal velocity $\sqrt{\gamma p/\rho}$ of the ions is included. The neutral flow is assumed to be supersonic with negligible thermal velocity. The average cross-section for neutral-ion interaction in the mixture of Huebner et al. is equal to $Q = 6.08 \cdot 10^{-15}$ cm^2. The elastic collisions also effect an energy transfer

$$\dot{E} = F \frac{2 m_i m_n}{(m_i + m_n)^2} \left(\frac{m_n v^2}{2} - \left(\frac{m_i u^2}{2} + \frac{1}{\gamma - 1} \frac{p}{N_i} \right) \right). \tag{3}$$

The last term on the right hand side is the average thermal energy of an ion. In most parts of the flow the average energy of the ions is larger than that of the neutrals. Therefore E is negative. Formulae (2) and (3) are obtained from the formulae for two-body collisions by averaging over the impact parameter.

We describe the interaction by the hydrodynamic or magneto-hydrodynamic equations with source terms, which represent the influence of the comet. The pick-up of ions is taken into account by a source term $\dot{M}$ in the continuity equation, the collisions with the neutral gas by the source terms $\dot{I}$ in the momentum and $\dot{E}$ in the energy equation given by equations (1) and (3).

The comet disturbs the solar wind in a rather large region, but as long as this disturbance is small, the solar wind remains a parallel flow. Let x,y,z be a system of coordinates centered at the comet with the z-axis pointing in the direction towards the sun. The velocity vector of the solar wind is then antialigned to the z-axis.

The stationary equations for such a parallel flow can easily be integrated along each streamline. One obtains equations for the density ρ, the velocity u and the pressure p at each point in dependence of the coordinate z.

$$\rho(z)\ u(z) = \rho_{sw}\ u_{sw} - \int_z^\infty \dot{M}\ dz \tag{4}$$

$$\rho(z)\ u^2(z) + p(z) = \rho_{sw}\ u_{sw}^2 + p_{sw} - \int_z^\infty \dot{I}\ dz \tag{5}$$

$$\left(\frac{\rho(z)u^2(z)}{2} + \frac{\gamma}{\gamma-1}\ p(z)\right) u(z) = \left(\frac{\rho_{sw}\ u_{sw}^2}{2} + \frac{\gamma}{\gamma-1}\ p_{sw}\right) u_{sw} - \int_z^\infty \dot{E} dz \tag{6}$$

Here, ρ_{sw}, u_{sw}, p_{sw} denote the density, velocity and ion pressure in the solar wind. Note that in view of the orientation of the coordinate system u and u_{sw} are negative. The integrals on the right hand side are the mass ΔM, the momentum ΔI and the energy ΔE already injected into the flow before it reached this point. The left hand sides are the flows of mass, momentum and energy defined by

$$M := \rho u, \qquad I := \rho u^2 + p, \qquad E := \left(\frac{\rho u^2}{2} + \frac{\gamma}{\gamma-1}\ p\right) u \tag{7}$$

This is an algebraic system of equations for ρ, u and p, which for given M, I, and E has a solution only if the inequality

$$\frac{2EM}{I^2} \leqslant \frac{\gamma^2}{\gamma^2-1} \tag{8}$$

is satisfied. We observe that on the right hand sides of (4), (5) and (6) the integrals are small compared to the first terms. Since the solar wind is hypersonic, p_{sw} is negligible compared to $\rho_{sw}u_{sw}^2$. In view of this, equations (4), (5) and (6) have a solution $\rho(z)$, $u(z)$ and $p(z)$ only as long as the added mass ΔM, momentum ΔI and energy ΔE satisfy

$$\frac{\Delta M}{|\rho_{sw}u_{sw}|} + 2\frac{\Delta I}{\rho_{sw}u_{sw}^2} + \frac{2\Delta E}{|\rho_{sw}u_{sw}^3|} \leqslant \frac{1}{\gamma^2-1} . \tag{9}$$

We use the ratio of specific heats $\gamma = 2$ and observe that for the Giacobini-Zinner model along the axis comet-sun the left hand side in (9) is less than 1/3 only for distances from the nucleus larger than 33 000 km. At this point the relative importance of the source terms can be deduced from

$$\frac{\Delta M}{|\rho_{sw}u_{sw}|} = .23 , \quad \frac{\Delta I}{\rho_{sw}u_{sw}^2} = .05 , \quad \frac{2\Delta E}{|\rho_{sw}u_{sw}^3|} = -.01 . \tag{10}$$

Before (8) is violated, a shock must occur, which allows the flow to become divergent and in this way to cope with the newly added material.

Now I report on an axisymmetric calculation. In this model we solve besides the hydrodynamic equations mentioned above also the continuity equations

$$\text{div}\, N_i \vec{u} = \dot{N}_i \tag{11}$$

for 45 different ions. The source terms on the right hand sides in (11) take into account the generation of ions by photo-processes or electron-impact from the spherically symmetric neutral gas cloud described above, but they include also the transformation of protons and ions by charge exchange with neutrals and the annihilation of ions by recombination with electrons. In our model we use a detailed chemistry with 90 different atoms, molecules and ions and 513 reactions. Many of these reactions generate or consume energy, mainly energy of the electrons. Therefore it is necessary to add an equation for the book-keeping of the energy of the electrons.

The number densities which result from this axisymmetric model along the comet sun axis are shown in figure 2a. They can be compared with the results for the undisturbed comet shown in figure 1b. Figure 2b shows that far away from the nucleus the comet effects mainly that the temperature and the pressure of the ions increase, and the Machnumber decreases. Finally a shock occurs at a distance of 50 000 km. This shock is weak (Ma~2). It compresses the flow by a factor of 1.7. Inside the shock the ion pressure remains almost constant in a large region. Approaching the nucleus, the mean molecular weight increases from 1 to about 22, since more heavy ions are added. The velocity decreases. Stagnation of the flow may be expected somewhat inside the inner edge of the domain of calculation at a distance of about 1000 km.

Figure 3b shows the amount of ion mass which is produced by the different ionization processes along the axis in dependence of the distance from the nucleus. Photo-ionisation is by far the most efficient process, followed by electron impact and charge exchange. A small fraction of the ions is removed by recombination.

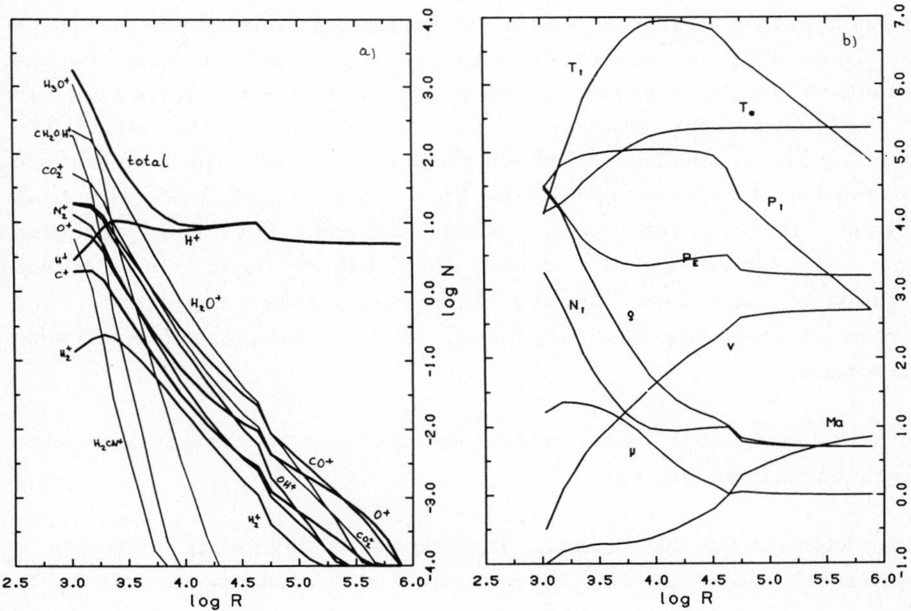

Fig.2. Solar wind-comet interaction. Values along the axis sun-comet. a) Number densities of some ions in cm^{-3}. b) Temperatures of ions T_i and electrons T_e in degree K, pressures of ions P_i and electrons P_e in 10^{-13} dyn/cm^2, velocity v in km/sec, Machnumber Ma, mass density ρ in m_p/cm^3, total number density N_i of ions in cm^{-3} and mean molecular weight μ.

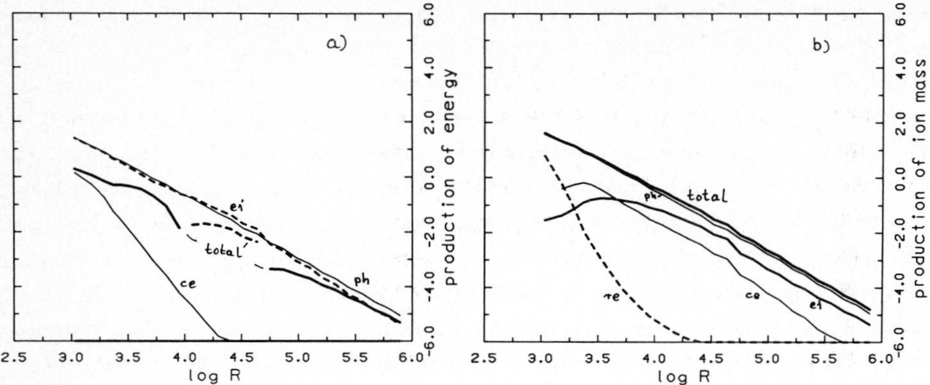

Fig.3. Generation of energy (a) in eV $s^{-1}cm^{-3}$ and ion mass (b) in m_p $s^{-1}cm^{-3}$ by photoprocesses (ph), electron impact (ei), charge exchange (ce) and recombination (re). Negative contributions are distinguished by dashed lines.

The contribution of the different processes to the energy balance along the axis is shown in figure 3a. Photo-ionisation provides the main source of energy. A comparable amount of energy is consumed by electron impact reactions. The efficiency of this cooling is proportional to the electron density which is suddenly changed by the shock compression. In this way the shock acts like a switch. In front of the shock the electrons are heated, behind the shock they are cooled. The temperature of the electrons decreases (Fig.2b). This reduces the efficiency of the electron impact reactions, which depends on the electron temperature. Therefore in the dense cold region near the nucleus electron impact reactions become less and less important, and recombination becomes the main opponent, which diminishes the effects of photoprocesses.

Twodimensional plots for this model as well as a comparison with ICE-data are contained in [4].

Let us now turn to an MHD model. The expected asymmetry requires a three-dimensional calculation. For reasons of economy we can only include part of the chemistry. As I have shown in figure 3 we don't loose very much if we consider only photo-ionisation. We use the numerical method as described in [5], with the following changes. We calculate two quadrants to describe inclined fields. We use a refined representation of the source terms based on the Huebner model and include neutral-ion collisions as described above. We also include more of the foreland into the calculation. The numerical method is capable to treat the shock front automatically.

The plots in figure 4 show the situation in a meridional plane cut parallel to the interplanetary magnetic field. At a distance of about 1 a.u. from the sun the field is inclined with respect to the velocity vector by 45 degrees. On one side the field is parallel to the shock and therefore compressed by the shock. On the other side it is nearly perpendicular and remains essentially unaffected by the shock transition. Therefore in figure 4a) the shock can be only recognized as a field compression in the right hand side of the plot. Near the nucleus there is a region where the field is amplified by a factor of about 4 over its interplanetary value to 35 nT. The field lines are draped around the nucleus. They are isochrones, which mark the flow at certain time intervals. The field lines in figure 4b follow each other

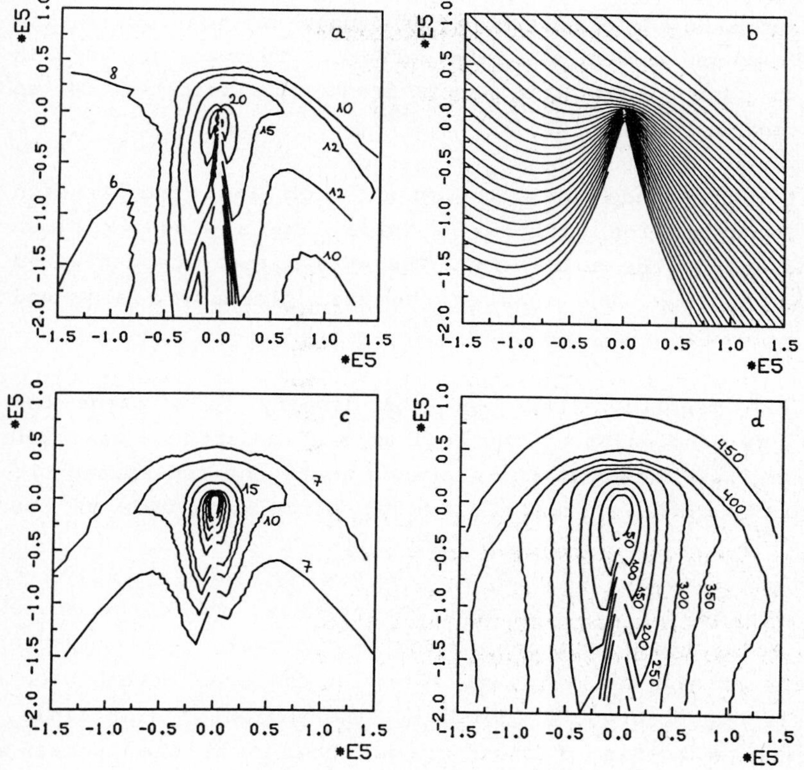

Fig.4. Cut through an MHD-model in a meridional plane parallel to the IMF. a) total field |B| in nT, b) some field lines, c) mass density in $m_p cm^{-3}$ (inner contours 20,50, 100,200,500,1000,...), d) velocity |u| in km/s.

at time intervals of 30 seconds. Using the picture of the frozen-in field lines one gets an intuitive feeling how these field lines act like ropes, dragging the cometary material into the tail. In fact the magnetic field transfers efficiently momentum from the bystreaming solar wind to the cometary tail.

In view of the fact, that this model was calculated with a numerical method which does not conserve magnetic flux automatically, one might worry about the possibility that field lines end somewhere at some artificial monopoles. But the picture with the beautifully draped field lines testifies the validity of the numerical solution.

In the solar wind the effect of the Lorentz forces is negligible. Even the amplified field behind the shock has little influence on the flow. Therefore in most parts the MHD solution looks very similar to a hydrodynamical solution. Nevertheless the magnetic field topology is to some extent reflected in the plasma flow. It causes a slight aberration of the tail, and imposes a dovetail like pattern. This can be seen in figure 4c and d, where also the shock front is indicated by an accumulation of contour lines.

The visible tail corresponds to a region of high ion concentration hence high mass density. The tail has a shape like a windsock blown away from the nucleus by the solar wind. The streamlines are deflected in passing by the nucleus. The ions in the tail plasma are slow and cool.

Figure 5 shows the magnetic field and the density in a plane cut through the tail, perpendicular to the tail axis at a distance of 7 800 behind the nucleus. This is about the distance where the ICE spaceprobe passed. These plots cover an area of 100 000 km on each side of the axis.

The shock front is clearly visible near the corners. The large scale asymmetry of the total field strength is due to the fact, that only the parallel component of the field is amplified in the shock transition. Near the tail axis one observes a dumb-bell like region of high field. This is essentially a bundle of field lines hung up at the nucleus. Because of the draping of the field lines (fig.4b) the field component parallel to the sun-comet axis, changes sign. This happens at the dotted line in fig.5a. As a consequence the two lobes of strong field in the tail have opposite polarity.

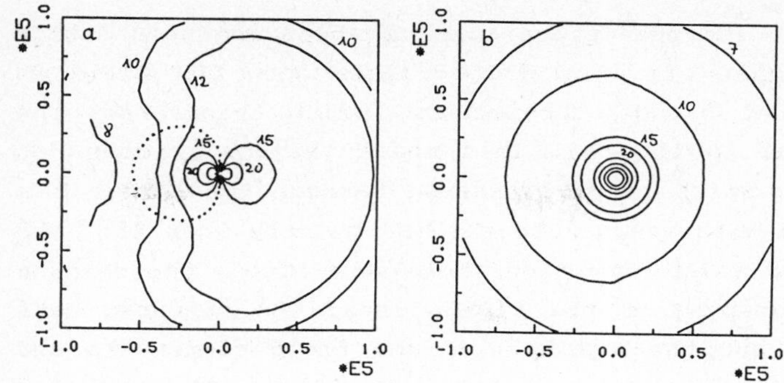

Fig.5. Cut through an MHD-model in a plane perpendicular to the tail axis. a) total field $|B|$ in nT. At the dotted line the field component perpendicular to this plane changes sign. b) mass density in $m_p cm^{-3}$.

In the density plot (fig.5b) the contour lines seem to be shifted somewhat to the left. This indicates the aberration of the tail caused by the magnetic field.

Finally, we compare this model with the data obtained from the ICE spacecraft during the encounter with comet Giacobini-Zinner. For the interpretation one must keep in mind, that the undisturbed solar wind needs ten minutes to transverse the whole region depicted in figure 4, but ICE needed four hours. Therefore the spacecraft measured not only spatial variations within the comet, but also temporal variations in the solar wind.

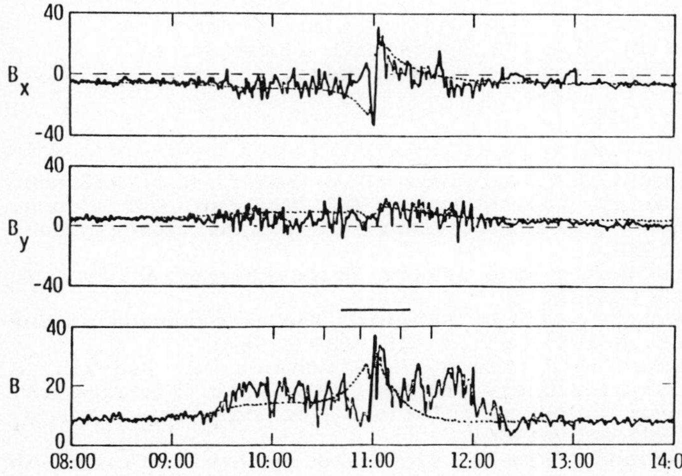

Fig.6. Magnetic field in nT measured by the ICE spacecraft (Smith et al. [7]) and field along a trajectory in an MHD-model (dotted lines).

Figure 6a shows what the ICE-spaceprobe told us about magnetic fields near comet Giacobini-Zinner [7]. The dotted lines show the field we obtained in our MHD model along a trajectory, which transverses the tail in a plane parallel to the IMF. This model trajectory does not match the ICE trajectory which probably passed through the comet in a plane which was inclined with respect to the IMF by more than 45°. The field component B_x in the solar direction shows very nicely the changes in sign caused by the draping of the field lines. The measured data show strong turbulence, but the averages of the field components and the velocity are rather well-represented by our model. The observed total field is systematically higher. This may be due to a magnetic field generated by the turbulence.

The MHD model differs in several respects from the hydrodynamical model, which was compared with ICE measurements in [4]. In figure 7 I have drawn the velocity profile in the MHD model along the trajectory described above, and the velocity profile measured by the ICE (Bame et al. [1]). Comparison shows reasonably good agreement of theory and experiment.

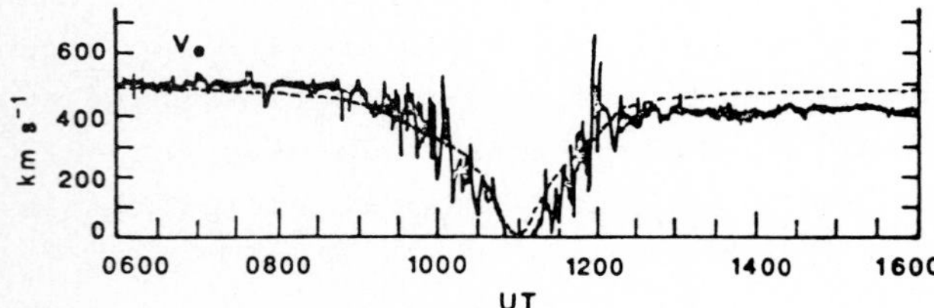

Fig.7. Velocity measured by the ICE spacecraft (Bame et al. [1]) and along a trajectory in an MHD model.

References

1. Bame, S.J., R.C. Anderson, J.R. Axbridge, D.N. Baker, W.C. Feldman, S.A. Fuselier, J.T. Gosling, D.J. McComas, M.F. Thomsen, D.T. Young, and R.D. Zwickl, Comet Giacobini-Zinner: A plasma description, Science 232, 356-360 (1986).
2. Biermann, L., Kometenschweife und solare Korpuskularstrahlung, Z. Astrophys., 29, 274-286 (1951).
3. Biermann, L., B. Brosowski, and H.U. Schmidt, The interaction of the solar wind with a comet, Solar Phys. 1, 254-284 (1967).
4. Boice, D.C., W.F. Huebner, J.J. Keady, H.U. Schmidt, R. Wegmann, A model for comet P/Giacobini-Zinner, Geophys. Res. Lett. (to appear).
5. Schmidt, H.U., R. Wegmann, MHD-calculations for cometary plasmas, Comp. Phys. Comm. 19, 309-326 (1980).
6. Schmidt, H.U., R. Wegmann, Plasma flow and magnetic fields in comets. In Comets (ed. Wilkening L.L.) 538-560, University of Arizona Press, Tucson, 1982.
7. Smith, E.J., B.T. Tsurutani, J.A. Slavin, D.E. Jones, G.L. Siscoe, D.A. Mendis, ICE encounter with Giacobini-Zinner: Magnetic field observations, Science 232, 382-385 (1986).

Part II

The Sun as a Star

Rotation and Magnetic Fields in the Sun

N.O. Weiss

Department of Applied Mathematics and Theoretical Physics,
University of Cambridge, Cambridge CB3 9EW, UK

Stellar magnetic fields exert torques which alter the distribution of angular momentum in a star. In the radiative interior of the Sun these torques tend to enforce uniform rotation and the existence of a rapidly rotating core would imply a poloidal field of less than 10^{-2}G. In the convective envelope magnetic fields generated by dynamo action produce torques which lead to torsional oscillations. A simple nonlinear model allows both multiply periodic and chaotic behaviour. This system demonstrates that both aperiodic magnetic cycles and the irregular modulation responsible for grand minima can be regarded as examples of deterministic chaos. This picture is consistent with the ^{14}C record but implies that the multiply periodic lamination of some Precambrian varves is not associated with the solar cycle.

1. Introduction

One characteristic feature of research at the Max Planck Institute for Astrophysics is the link between solar and stellar physics, which existed long before it became fashionable to talk about the solar-stellar connection. Another is the close relationship between theory and observations. I shall discuss two problems involving interactions between magnetic fields and rotation, both stimulated by recent observations. In each case, solar behaviour can only be explained by reference to other stars; and in each case there seems to be a conflict between the observations and our theoretical expectations.

The first problem is posed by recent measurements of rotational splitting of the frequencies of solar five-minute oscillations, which suggest that the radiative core of the Sun may be rotating at twice the surface rate (Duvall & Harvey 1984). Is the associated shear in angular velocity compatible with the presence of a significant poloidal field in the radiative zone, and what field strength might we expect to find there? In a perfectly conducting gas, magnetic torques alter the pattern of differential rotation on a timescale comparable with the Alfvénic transit time; if this time is comparable with the age of the Sun, the poloidal field cannot be stronger than a few microgauss. L.Mestel and I have tried to estimate the effect of instabilities on the toroidal field, and we find that this upper bound could be raised to 3×10^{-4}G or even, perhaps, to 3×10^{-2}G. We have also estimated the field that might have been left in the radiative zone as the convective zone retreated during the Sun's approach to the main sequence. This poloidal field might be as low as 5×10^{-2}G but is more likely to be in the gauss to Kilogauss range. Thus the expected field strengths are scarcely compatible with the survival of a strong rotational shear within the radiative zone.

The second issue concerns the apparent aperiodicity of the solar cycle. There is no viable alternative to a dynamo as the source of the Sun's oscillatory magnetic field. Numerical simulations have confirmed that convective motion in a rotating system can support a cyclic magnetic field (while casting doubt on the assumptions incorporated in specific models, tuned to reproduce the detailed pattern of the solar cycle). In a nonlinear regime, magnetic torques affect rotation in the convective zone, producing (for instance) torsional waves with twice the frequency of the magnetic cycle. F.Cattaneo, C.A.Jones and I incorporated this effect in a truncated model system and found solutions that exhibited both aperiodic cycles and a long period modulation of activity, resembling that associated with the Maunder minimum. In the solar dynamo there are many nonlinear effects and it is not clear which of them limits growth of the magnetic field. Nevertheless, our toy system shows that the observed aperiodicity of the solar cycle can be explained as an example of deterministic chaos without invoking any stochastic effects. Unfortunately, the historical record of solar activity only covers the period since telescopes were invented but this record matches that derived from anomalies in the production rate of ^{14}C in the atmosphere which can be extended back over the last 7000 years. There is, however, another remarkable data set which (it has been argued) preserves a record of solar activity 700 million years ago. The layered sediments of the Elatina formation in S. Australia show a predominant periodicity of about 12 years (Williams, 1981, 1985) and, since 20 000 annual layers are preserved, this time series can be analysed in detail. Frequency analysis shows a line spectrum, corresponding to multiply periodic behaviour. I shall argue, regretfully, that such a pattern is not compatible with the aperiodic variation of magnetic activity in the present Sun.

2. Rotation and magnetic activity in the Sun

Helioseismology makes it possible to determine the angular velocity $\Omega(r,\theta)$, within the Sun by measuring the rotational splitting of the frequencies of p-modes with surface displacements corresponding to spherical harmonics $Y_{\ell}^{m}(\theta,\phi)$, referred to spherical polar co-ordinates (r,θ,ϕ). Duvall & Harvey (1984) used sectoral harmonics ($m=\ell$) to determine the variation with radius of the angular velocity at the equator (Duvall et al. 1984). Subsequently, the latitudinal variation of Ω was explored by Brown (1985), Libbrecht (1986) and Duvall, Harvey & Pomerantz (1986). These results can be assembled to give a general picture of the variation of Ω in the Sun that is broadly consistent with observations and with certain theoretical preconceptions (Rosner & Weiss 1985). Surfaces of constant Ω are shown schematically in Figure 1.

At the equator Ω first rises slightly above its surface value and then falls off slowly with depth in the convective zone, decreasing further in the outer part of the radiative zone. Numerical simulations of convection in a rotating system lead one to expect that $\partial\Omega/\partial r$ should be positive in the convective zone, with a tendency for Ω to be constant on cylindrical surfaces at low latitudes (Gilman 1979; Glatzmaier 1985). The variation of Ω with latitude observed at the surface apparently persists through the convective zone (Duvall et al. 1986) but probably diminishes with depth in the radiative interior (Brown 1985; Libbrecht 1986). Indeed, one would expect Ω to be a function of r only in the radiative core, since shear instabilities could otherwise lead to mixing on spherical surfaces (Zahn 1983). By $r=0.5R_{\odot}$, therefore, one would expect the angular velocity to be independent of latitude, with a value that is some weighted average of $\Omega(r_c,\theta)$, where r_c is the radius at the base of the convective envelope. Frequency splitting of 5-minute oscillations indicates two further features in the radiative zone. Modes with degree $\ell=11$ show enhanced splitting, corresponding to more rapid rotation, in measurements made by Duvall & Harvey (1984) and Brown (1985) but not

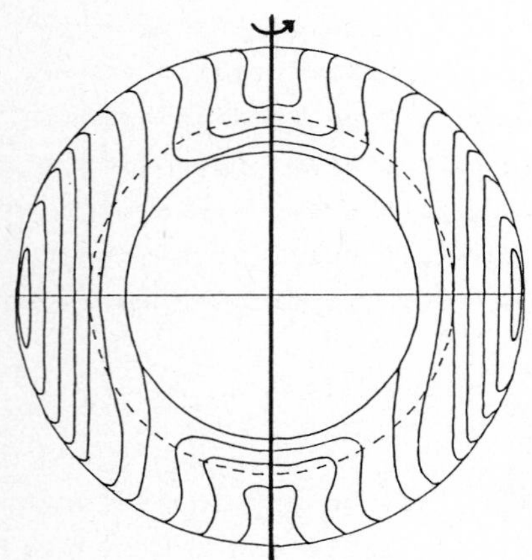

Figure 1. Sketch showing surfaces of constant angular velocity in the Sun, after Rosner & Weiss (1985). The broken line shows the base of the convection zone.

in those of Libbrecht (1986). On theoretical grounds, it is hard to accept the reality of an isolated maximum in $\Omega(r)$ and it seems more likely that the observed splitting has some other cause (Rosner & Weiss 1985). The second and more remarkable feature is an enhanced splitting of modes with $\ell=1,2$, which implies that Ω rises to about twice its surface value at $r\approx 0.2R_\odot$. The existence of such a rapidly rotating core has important implications for the rotational and magnetic history of the Sun.

From observations of stars in clusters it is possible to determine the evolution of the surface rotation rate of a star of solar mass on the main sequence. G stars rotate very rapidly, with periods around 12 hours, when they arrive on the main sequence but slow down rapidly so that the periods rise to 3d within 3×10^7yr (Stauffer et al. 1984, 1985); thereafter, Ω decays more gradually, decreasing by an order of magnitude over 5×10^9 yr. Angular momentum is lost by magnetic braking, which leads to spindown at a rate depending on Ω, since the strength of the dynamo-produced field is itself a function of Ω. Thus magnetic activity and rotation decrease monotonically with age in a main sequence star. In particular, the Sun rotated at twice its present rate about 3×10^9 yr ago. Since magnetic torques are efficient at eliminating differential rotation, one is led to ask whether the existence of a rapidly rotating core is compatible with the presence of a significant magnetic field in the radiative zone (Mestel & Weiss 1986).

3. Magnetic fields and differential rotation

It is well known that in a perfectly conducting star with an axisymmetric magnetic field the angular velocity must be constant along field lines in a steady state. If Ω varies, the shear gives rise to torsional waves which travel with the Alfvén speed $v_A=|B_P|/(4\pi\rho)^{\frac{1}{2}}$, where $\underset{\sim}{B}_P$ is the poloidal component of the field. Thus the appropriate timescale for magnetic torques to alter the distribution of angular momentum is $\tau_A=D/v_A$, where D is the scale of the shear (assumed small). Alfvén waves will be dissipated within the radiative core and scattered in the outer parts of the radiative zone, which are dynamically coupled to the convective envelope. Hence the braking time, τ_b, for differential rotation is just the Alfvén time τ_A. If $\tau_b\geqq 3\times10^9$ yr we require that $B_P\leqq 3\times10^{-6}$G: but it seems absurd that the magnetic field in a star should have a value typical of the interstellar medium.

This argument depends on the assumption that the toroidal field, B_ϕ, satisfies the induction equation

$$\frac{\partial B_\phi}{\partial t} = r\sin\theta\ \underset{\sim}{B}_P \cdot \nabla\Omega \qquad (3.1)$$

and grows linearly with time. Can this growth be limited by instabilities, so that

$$\left|\frac{B_\phi}{B_p}\right| \approx \left(\frac{r\sin\theta}{D}\right) \Omega\tau \qquad (3.2)$$

for some τ (Mestel & Weiss 1986)? A predominantly toroidal field is dynamically unstable (c.f. Pitts & Tayler 1985 and references therein). In the neighbourhood of the magnetic axis the configuration resembles a pinched discharge, which is liable to kink and sausage instabilities with a characteristic growth time $\tau = D/w_A$, where $w_A = |B_\phi|/(4\pi\rho)^{\frac{1}{2}}$. Since $|B_\phi| >> |B_p|$ we can obtain an upper bound to the poloidal field strength in the radiative core by using this value of τ in (3.2), whence it follows that if $\tau_b \gtrsim 3\times10^9$ yr then $|B_p| \lesssim 3\times10^{-2}$G. This argument is too optimistic, however, for instability is inhibited by gravitational stratification as well as rapid rotation. Pitts & Tayler (1985) estimate that the growth time is increased by a factor (v_s/w_a), where the sound speed $v_s >> w_A$. In that case, we find that $|B_p| \lesssim 3\times10^{-4}$G. There is the alternative possibility that resistive instabilities might lead to rapid reconnection near the equator but it seems probable that the tearing mode will saturate at low amplitudes (eg. Priest 1985).

What field strength should one expect within the radiative core? If magnetic flux is trapped within a collapsing protostar the field could be as high as 10^8 gauss but it is generally supposed that a star loses this primeval flux during the Hayashi phase and arrives on the main sequence with a magnetic field well below the virial limit. By this stage, magnetic activity is produced by dynamo action in the convective envelope, though it is not clear whether the dynamo is steady or oscillatory. As the star evolves towards the main sequence the convective zone retreats outwards, leaving a fossil field in the radiative core. Estimates of the residual field strength vary from 10^3G, if the dynamo is stationary (Schüssler 1975; Parker 1981) to 5×10^{-2}G for an oscillatory dynamo (Mestel & Weiss 1986). B_p is probably in the gauss to Kilogauss range and, on balance, it seems unlikely that the field can be small enough for differential rotation to survive. If the deduced value of Ω in the radiative core is correct, then either Ω is unsteady (Gough 1985) or else the core is magnetically isolated from the rest of the star (Rosner & Weiss 1985; Roxburgh 1986).

More precise measurements of the splitting of modes with $\ell=1$ will eventually allow us to determine whether the core spins at twice the surface rate - or whether the whole Sun rotates more or less uniformly after all.

4. Modulation of cyclic activity in stars

Stellar magnetic activity is closely correlated with rotation (Baliunas & Vaughan 1985) and hydromagnetic dynamos rely both on differential rotation and on helicity caused by Coriolis forces. Cyclic variation in activity seem to be typical of slowly rotating stars with deep convection zones, like the Sun. Over the last 270 years solar activity has oscillated irregularly, with a well-defined average period of about 11 years; since the field reverses at sunspot minimum the corresponding magnetic cycle has a 22-year period. This cyclic pattern is modulated on a longer timescale, with grand minima when activity almost disappears. The best example is the Maunder minimum of the 17th century, when the incidence of sunspots was drastically reduced. This was noticed by contemporary astronomers, such as Hevelius, whose competence was well established, and even referred to by Andrew Marvell in a poem (Weiss & Weiss 1979). This grand minimum, lasting for 70 years, was rediscovered by Spörer (1889) and investigated by Maunder and by Eddy(1976). Spörer noticed not only the reduction in activity but also the changed distribution of sunspots in latitude. Prior to the Maunder minimum, cyclic activity occurred as now: sunspot zones migrated towards the equator and spots reappeared at higher latitudes (around 25°) after sunspot minima (e.g. that of 1619). Yet of about 60 spots recorded between 1671 and 1713 only 7 were at latitudes greater than 13° (and almost all were in the southern hemisphere). Not till 1714 did spots appear at higher latitudes and in the northern hemisphere. This anomalous distribution in latitude confirms the reality of the Maunder minimum. (It is also worth noting that during the "little minimum" of 1795-1799 the only spots recorded were at low latitudes.) Apparently grand minima are associated with a reduction in activity resembling a postponement of the next full cycle, though it cannot be established whether cyclic behaviour persists at a reduced level or with a different period.

Fortunately, it is possible to extend the record of grand minima backwards for about 7000 years by using anomalies in the ^{14}C production rate, which can be precisely dated from tree rings. The physical mechanism is clear: ^{14}C is formed by neutrons produced by galactic cosmic rays, whose incidence is affected by magnetic fields in the solar wind and therefore varies with the solar cycle. Since ^{14}C atoms remain in the atmosphere for about 50 years only the envelope of the solar activity curve can be determined by this method. This envelope agrees well with that derived from observations since the time of Galileo and has been carried back with great precision over the last millennium, as shown in Figure 2(a) (Stuiver & Quay 1980; Stuiver & Grootes 1981). In particular, it shows the Maunder minimum, which has also been detected in measurements of ^{10}Be abundances in ice-cores (Raisbeck et al. 1981). In addition, there are several earlier episodes of modulation, which seem consistent with the scanty record of early sunspot and auroral observations. Going further back, similar episodes recur, for instance, in the ^{14}C record over a 400 year interval around 3500 BC (Stuiver 1980) though these modulations are significantly smaller than the variation caused by gradual changes in the Earth's magnetic field (Eddy 1976).

Two features of the ^{14}C record must be emphasized. First, many of the grand minima appear surprisingly similar. Stuiver (1980) demonstrates the resemblance between six intervals ranging from 3650 BC to 1610 AD and, in particular, the close correspondence between the intervals 970-1250 and 1610-1860. He also notes the occasional appearance of a different, longer interruption as, for instance, in the interval 1570-1610. Secondly, the modulation is not periodic. Grand minima were most severe around 1670, 1450, 1330, 1040, 900 and 790 AD and, further back, around 3320, 3500 and 3620 BC (Stuiver 1980; Stuiver & Grootes 1981), giving intervals of 220, 110, 290, 140, 110, 180 and 120 years respectively. The frequency spectrum shows broad peaks centred on periods around 130 years and 200 years, as one might expect. These results impose constraints on theories of the solar cycle: they must be able to explain an irregular cyclic pattern of modulations, with recurrent episodes that are similar but aperiodic. As we shall see, such behaviour is characteristic of a class of nonlinear oscillators and can be simulated with simple dynamo models of stellar magnetic cycles.

5. Nonlinear dynamos and deterministic chaos

Hydromagnetic dynamo theory still provides the only viable explanation for cyclic magnetic activity in stars like the Sun (cf. Parker 1979; Cowling 1981). In the past, simplified models have been tuned to reproduce the observed details of the solar cycle and there was some embarrassment when it turned out that these models were based on assumptions that were no longer appropriate (e.g. $\partial\Omega/\partial r<0$ in the convective zone). Self-consistent nonlinear computations have, however, confirmed that cyclic fields can be generated in a convecting layer confined between concentric spherical shells (Gilman 1983; Glatzmaier 1985). In these calculations, toroidal fields migrate towards the pole instead of towards the equator, as observed on the Sun. The process works but the details are wrong. This merely suggests that the detailed physics of a stellar dynamo is not yet adequately represented in the numerical models. Indeed, it is not obvious where the dynamo is located, though it seems likely that magnetic fields are generated near the base of the convective zone, become unstable through magnetic buoyancy and float upwards to emerge in active regions.

In a nonlinear dynamo model the induction equation

$$\frac{\partial \mathbf{B}}{\partial t} = \text{curl } (\mathbf{u}\times\mathbf{B}) + \eta\nabla^2\mathbf{B} \tag{5.1}$$

must be solved for the magnetic field $\mathbf{B}$, where η is the magnetic diffusivity and the velocity $\mathbf{u}$ is obtained by solving the nonlinear equation of motion. Much of the work in this subject has been concerned with mean field dynamo models. We suppose that the induction equation can be averaged azimuthally and that the effects of turbulence can be parametrized; this procedure can be justified in certain circumstances, which do not apply to the Sun. Then we set

$$\mathbf{B} = \mathbf{B}_P + B_\phi\hat{\boldsymbol{\phi}}, \qquad \mathbf{B}_P = \text{curl } (A\hat{\boldsymbol{\phi}}) \tag{5.2}$$

and the poloidal and toroidal components of (5.1) yield the linear (kinematic) dynamo equations

$$\begin{aligned} \frac{\partial A}{\partial t} &= \alpha B_\phi + \eta\left(\nabla^2 - \frac{1}{r^2\sin^2\theta}\right)A, \\ \frac{\partial B_\phi}{\partial t} &= r\sin\theta\mathbf{B}_P.\nabla\Omega + \eta\left(\nabla^2 - \frac{1}{r^2\sin^2\theta}\right)B_\phi, \end{aligned} \tag{5.3}$$

respectively. Here α is proportional to the mean helicity $\langle\mathbf{u}.\ \text{curl}\ \mathbf{u}\rangle$ and describes the generation of poloidal flux by gyrotropic motion, while η is now a turbulent diffusivity. The behaviour of solutions of (5.3) depends on a stability parameter, the dynamo number

$$D = \frac{\alpha\Omega' d^4}{\eta^2}, \tag{5.4}$$

where d is an appropriate length scale and Ω' is the angular velocity gradient. The trivial solution ($\mathbf{B}$=0) of (5.3) undergoes an oscillatory (Hopf) bifurcation and loses stability when $D=D_{crit}$. Thereafter, solutions to the linear problem give dynamo waves which propagate towards the equator if Ω increases inwards.

Nonlinear dynamos must take account of the effect of the magnetic field on the motion. In particular, magnetic torques affect differential rotation, giving rise to a fluctuating toroidal velocity v_ϕ that satisfies an equation of the form

$$\frac{\partial v_\phi}{\partial t} = \frac{1}{\mu_0\rho}\left[\text{curl } (B\hat{\boldsymbol{\phi}}) \times \mathbf{B}_P.\hat{\boldsymbol{\phi}}\right] + \nu\left(\nabla^2 - \frac{1}{r^2\sin^2\theta}\right)v_\phi , \tag{5.5}$$

where ν is a (turbulent) viscous diffusivity. Since the Lorenz force is quadratic in B we expect torsional waves with twice the frequency of the magnetic cycle. Such waves have been observed (LaBonte & Howard 1982) and appear consistent with predictions based on simple dynamo models (Schüssler 1981; Yoshimura 1981).

One way of investigating nonlinear behaviour is to seek the simplest system that captures the essential physics of a stellar dynamo. Suppose we consider plane dynamo waves, referred to cartesian axes such that

$$\underset{\sim}{B} = (0,\ \tilde{B},\ \partial\tilde{A}/\partial x), \qquad \underset{\sim}{v} = (0,\ v(z),\ 0), \tag{5.6}$$

where $\tilde{A}(x,t)=A(t)\exp(ikx)$ and $\tilde{B}(x,t)=B(t)\exp(ikx)$. Then (5.3) can be replaced by the linear equations (in dimensionless form)

$$\dot{A} = 2DB - A\ , \qquad \dot{B} = iA - B\ , \tag{5.7}$$

where the dynamo number$_n$

$$D = \alpha v'/2\eta^2 k^2\ . \tag{5.8}$$

Then there is an oscillatory bifurcation at $D=1$ and the trivial solution ($A=B=0$) is unstable to oscillatory modes (waves) for $D>1$ (Parker 1955). The corresponding nonlinear system is obtained by supposing that

$$v' = 1 + \omega(t)\ \exp\ (2ikx). \tag{5.9}$$

so that (5.3) and (5.5) are replaced by the sixth order complex system

$$\begin{aligned} \dot{A} &= 2DB - A, \\ \dot{B} &= iA - \tfrac{1}{2}iA^*\omega - B, \\ \dot{\omega} &= -iAB - \nu\omega. \end{aligned} \tag{5.10}$$

This system possesses solutions that show three successive oscillatory bifurcations, leading to multiply periodic solutions (quasiperiodic in mathematical terminology) with trajectories that lie on a 3-torus in phase space. Subsequently there is a transition, via period-doubling, to a chaotic regime in which the toroidal field B oscillates aperiodically with an amplitude that is modulated irregularly on a longer timescale (Weiss, Cattaneo & Jones 1985; Weiss 1985). This modulation produces episodes of reduced activity that resemble grand minima in the record of solar activity, as illustrated in Figure 2(b).

What do these results tell us about stellar activity? The toy system (5.10) mimics the behaviour of the solar cycle but it cannot represent the full nonlinear physics or the complicated spatial structure of a real dynamo. We learn, however, that episodic modulation can be found in simple systems, where it is associated with the break-up of tori to which trajectories in phase space are attracted. This can be illustrated with yet simpler systems. Consider, for instance, the third-order system

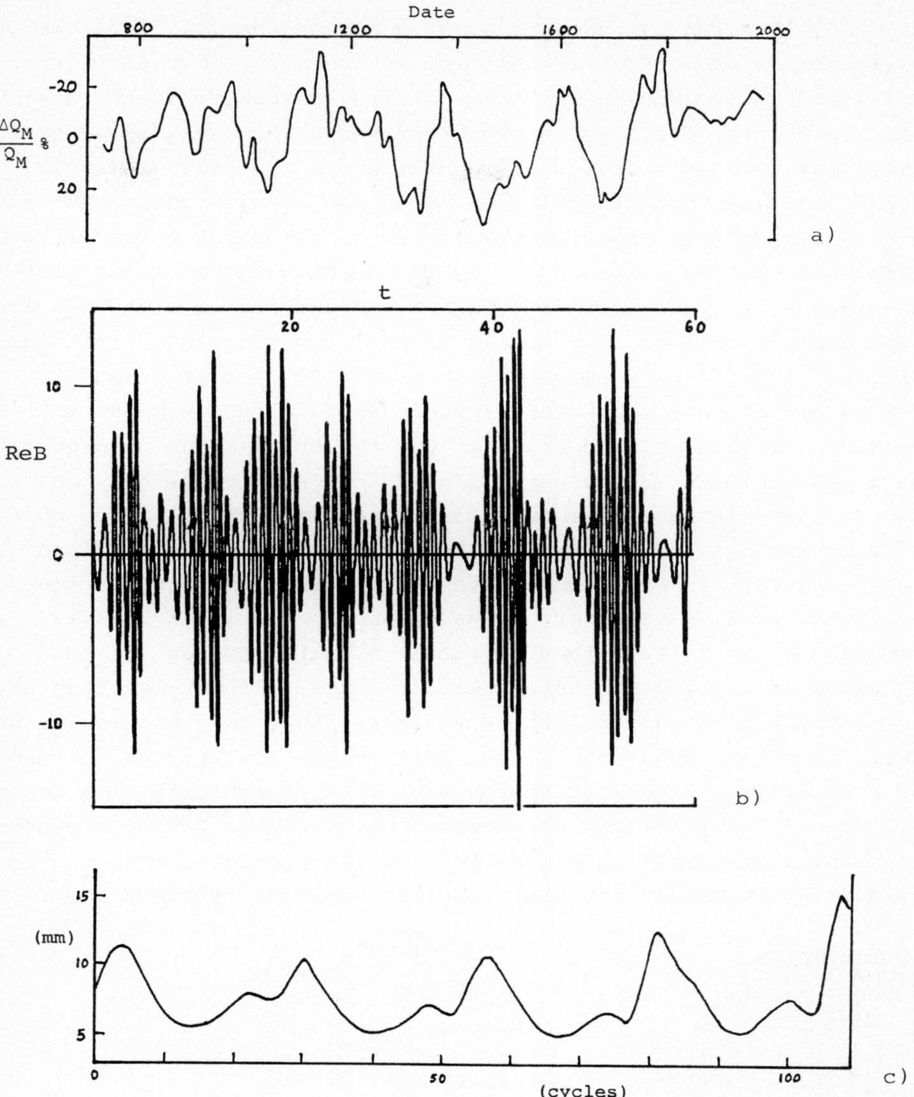

Figure 2. Modulation of solar activity. (a) Departure of the ^{14}C production rate from its mean value over the last millennium, after Stuiver & Grootes (1981). (b) Behaviour of a chaotic dynamo model, showing Re B as a function of time, after Weiss et al. (1985). (c) Smoothed record of thickness of varve cycles, after Williams (1985).

$$\dot{s} = \tfrac{1}{2}sz, \qquad \dot{\phi} = 15,$$
$$\dot{z} = \mu + z(1 - {}^{1}\!/_{3}z^2) + s^2(1 + 0.7s\cos\phi + 0.5z), \tag{5.11}$$

referred to cylindrical polar co-ordinates (s, ϕ, z) (Langford 1983). For $\mu=1$ trajectories are attracted to a 2-torus that resemble a pinecone with a hole drilled through its axis, as shown in Fig. 3(a). For $\mu=0.7$, trajectories spend most of the time in the neighbourhood of the z-axis and wind rapidly around the outer surface of the pinecone, as in Fig. 3(b). A plot of r(t) shows intermittent bursts of activity separated by prolonged inactive episodes. This modulation is typically periodic, though solutions become chaotic as $\mu \to 2/3$. So these simple models can help us to understand the observed modulation of the solar cycle. If the torus survives as a "ghost attractor" in the chaotic regime we may expect to see modulation of cyclic activity with a characteristic timescale and an envelope that is similar (but not identical) for successive episodes. The duration of a grand minimum cannot be predicted, for it depends critically on how close a trajectory approaches to a saddle-point in phase space. In particular, the phase of the cycles will not be preserved through a grand minimum. (The sunspot minimum of 1619 is the only one before the Maunder minimum whose date can be reliably established and - for what it is worth - it is out of phase with the minima of 1713 and 1724.) Such behaviour is consistent with the observed patterns in the ^{14}C record, described in the previous section.

So we are led to believe that aperiodic magnetic cycles in stars can be regarded as examples of deterministic chaos, without invoking any intrinsic stochasticity (Jones 1985; Ruzmaikin 1986). More precisely, we assert that the global behaviour of the Sun's magnetic field can be described by a chaotic (strange) attractor with some resemblance to a torus. If this is strongly attracting, the overall pattern will not be too sensitive to details of small-scale behaviour, which provide a weak stochastic element that enhances the basic unpredictability of the system.

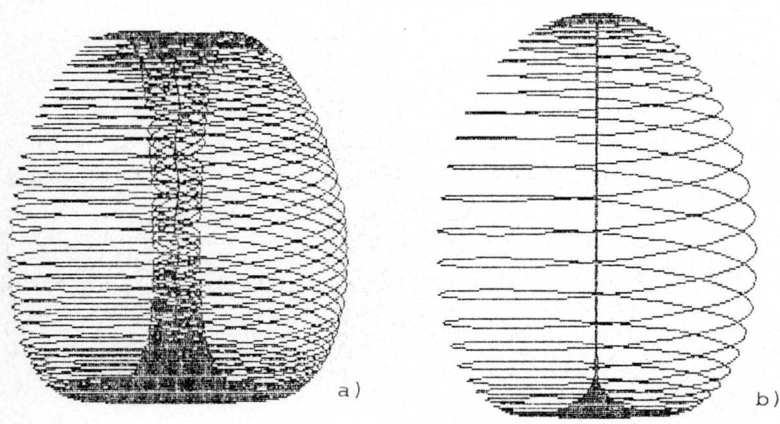

Figure 3. Trajectories on a 2-torus. Solutions of (5.11) for (a) $\mu=1$ and (b) $\mu=0.7$, showing trajectories projected onto the sz-plane.

In that case, we may indeed be able to discuss the mean field without becoming involved in details of individual active regions (just as climatic change can be distinguished from the weather).

6. Multiply periodic climatic oscillations

The picture outlined above seems consistent and - I hope - convincing but there exists an important and wholly different data set which leads to rather different conclusions. The varved sediments of the Elatina formation in South Australia have been described by G.E.Williams in a remarkable series of papers (Williams 1981, 1985; Williams & Sonett 1985). These Precambrian deposits were laid down in a glacial lake 7×10^8 years ago and 20000 "annual" layers are preserved. The layer thicknesses vary cyclically, with a mean period of about 12.0 years. In order to study modulation of this cyclic pattern it is convenient to measure the combined thicknesses of the layers that make up successive cycles. There is an obvious tendency for adjacent cycles to alternate in thickness and there is a long term modulation, shown in the smoothed curve of Figure 2(c), that repeats with surprising regularity and a mean period of 314 years. Correspondingly, the frequency spectrum shows sharp peaks at periods 2.1 and 26.2 times the mean cycle period. The sharpness of these peaks implies that the modulation is controlled by some multiply periodic (quasiperiodic) phenomenon.

The 12-year period, the drifting saw-tooth modulation (corresponding to a frequency close to twice that of the cycle) and the long-term modulation all remind one of solar activity. Indeed, the resemblance between the two patterns immediately suggests that these Precambrian varves may preserve a record of the solar cycle 700 million years ago. Such a time-series would be immensely valuable and would provide more information about the long-term behaviour of magnetic activity in the Sun than is available from any recent record. Wilson has argued forcefully in favour of this interpretation but one should nevertheless be cautious before accepting it.

The cyclic lamination must have been produced by climatic fluctuations. The main question is whether these have a terrestrial or a solar origin. Unfortunately, there is no clear mechanism (such as exists for ^{14}C production) for climatic forcing by solar activity. Since the time of William Herschel there have been attempts to establish a link between the solar cycle and climatic variations (not to mention trade cycles, vintage wines etc.). It is not difficult to pick out climatic oscillations with periods of around 11 or 22 years and there is evidence of such cycles in glacial deposits; but it has never been shown that such variations maintain their phase with respect to the solar cycle. It seems more likely that the climatic system processes many oscillatory modes with periods of 10-12 or 20-25 years and that these are easily excited. Perhaps there is a weak solar influence that can keep these oscillations synchronized with the activity cycle for a finite interval but its existence cannot easily be demonstrated.

There are certainly climatic variations with a characteristic timescale of several hundred years, of which the best known is the "Little Ice Age" that began around 1600 and continued for 250 years. Similar modulations have been noticed in other laminated deposits: for example, Permian evaporites from New Mexico and Texas provide a record covering more than 200,000 years and different segments yield periods ranging from 100 to 400 years (Anderson 1982). Detailed studies of glacial movements during the "Little Ice Age" show little correlation with the incidence of sunspots. Glaciers began to surge forwards around 1590 and there were particularly cold episodes before the Maunder minimum, in 1600-1610 and 1643-44 (Le Roy Ladurie 1972). After a warmer period in the eighteenth century, further cold episodes occurred in 1818-20 and 1850-55; since then the glaciers have been retreating. Stuiver (1980) compared the ^{14}C record with a number of climatic indicators and concluded that there was no significant correlation between climatic changes and solar variability over the last millennium. To be sure, the Elatina deposits were laid down 700 million years ago, during a severe ice age which affected Australia although it was at a low paleomagnetic latitude, and the climate might have been more susceptible to solar forcing. Williams (1985) suggests that the lower oxygen content of the atmosphere might have made the climate more responsive to changes in ozone concentration caused by variation in solar ultraviolet emission. Still, the available evidence is insufficient on its own to establish any links between climatic variations and the envelope of the activity cycle.

Such a link would be required if it could be demonstrated that the Elatina record was statistically similar to that of solar activity as preserved in ^{14}C abundances. There are obvious similarities and significant differences. Since 1700 the solar cycle has shown a well-defined mean period of 11.1 years, though individual cycles have ranged from 9 to 14 years. There are hints that periods may have been longer during the Maunder minimum but this cannot be confirmed; on theoretical grounds, one would expect a different frequency (determined by the eigenvalues of the linear problem) during a grand minimum (Weiss et al. 1984). Activity cycles have been recognized in 13 slowly rotating main sequence stars and this limited sample suggests that the cycle period increases with increasing rotation period. Since the Sun rotated more rapidly 7×10^8 years ago, the cycle period should have been 3-10% less than it is now (Noyes, Weiss & Vaughan 1984). Williams (1981) estimated a mean period of 11.2 years for the most prominent cycles in the varve sequence, which increases to 12.0 years when averaged over the 300 year cycle. There are signs of disagreement between the two periods but the astronomical data are too shaky for this incompatibility to be conclusive.

The major difference between the ^{14}C record in Figure 2(a) and the smoothed curve in Figure 2(c) is that the former seems aperiodic while the latter is close to being strictly periodic. If variations in varve thickness reflect changes in climate controlled by solar activity then the modulation of magnetic acitvity must be periodic. Of course this contradicts the interpretation put forward in the previous section. It is possible that the dynamo behaved differently in the late Precambrian but it seems unlikely that a relatively small change in rotation rate would produce a different pattern of convection (cf. Knobloch, Rosner & Weiss 1981). Dynamical systems do show narrow windows where periodic behaviour reappears as a parameter is increased but the probability of finding such a window is usually extremely small. If we accept that the ^{14}C record provides a faithful description of the modulation of solar activity then the appearance of grand minima over the last 1000 years, with intervals when magnetic activity almost disappears,is incompatible with the properties of the Elatina sequence.

The varves preserve a unique record of climatic variation which deserves careful analysis in its own right. If the modulation is not associated with magnetic activity then some other mechanism must be sought. The regularity of the variations in Figure 2(c) argues in favour of a non-terrestrial origin. It has been shown that precession and changes in the tilt and eccentricity of the earth's orbit produce long-term climatic modulations with timescales of 19,000 to 90,000 years. This implies that the climate can respond to a weak periodic or aperiodic forcing. The Sun remains the most obvious source for the Elatina cycle with a period of 314 years. So I conclude by speculating that there may be luminosity variations (or even a new solar oscillation) with this period. Unfortunately it would take several centuries for such a prediction to be verified.

Acknowledgements

I am grateful to George Williams for corresponding with me about his work and to Jack Eddy, Douglas Gough and Nicholas Shackleton for discussions, though the views expressed here are my own. Over the last 23 years I have benefitted from innumerable discussions with Rudolf Kippenhahn and Hermann Schmidt, which have led to fruitful and enjoyable collaboration. I would like to express my appreciation of the pleasure this has brought me.

References

Anderson, R.Y. 1982, J.Geophys.Res. 87, 7825.
Baliunas, S.L. & Vaughan, A.H. 1985, Ann.Rev.Astron.Astrophys. 23, 379.
Brown, T.M. 1985, Nature 317, 591.
Cowling, T.G. 1981, Ann.Rev.Astron.Astrophys. 19, 115.
Duvall, T.L., Dziembowski, W.A., Goode, P.R., Gough, D.O., Harvey, J.W. & Leibacher, J.W. 1984, Nature 310, 22.
Duvall, T.L. & Harvey, J.W. 1984, Nature 310, 19.
Duvall, T.L., Harvey, J.W. & Pomerantz, M.A. 1986, Nature 321, 500.
Eddy, J.A. 1976, Science 286, 1189.
Gilman, P.A. 1979, Astrophys.J. 231, 284.
Gilman, P.A. 1983, Astrophys.J.Suppl.Ser. 53, 243.
Glatzmaier, G.A. 1985, Astrophys.J. 291, 300.
Gough, D.O. 1985, in Future missions in solar, heliospheric and space plasma physics, ed. E.Rolfe, p.183, ESA SP-325, Noordwijk.
Jones, C.A. 1985, in The Hydromagnetics of the Sun, ed. T.D.Guyenne, p.91, ESA SP-220, Noordwijk.
Knobloch, E., Rosner, R. & Weiss, N.O. 1981, Mon.Not.Roy.Astron.Soc. 197, 45P.
LaBonte, B.J. & Howard, R.F. 1982, Solar Phys. 75, 161.
Langford, W.F. 1983, in Nonlinear dynamics and turbulence, ed. G.I.Barenblatt, G.Iooss & D.D.Joseph, p.215, Pitman, London.
Le Roy Ladurie, E. 1972, Times of feast, times of famine, Allen & Unwin, London.
Libbrecht, K.G. 1986, Nature 319, 753.
Mestel, L. & Weiss, N.O. 1986, Mon.Not.Roy.Astron.Soc., submitted.
Noyes, R.W., Weiss, N.O. & Vaughan, A.H. 1984, Astrophys.J. 287, 769.
Parker, E.N. 1955, Astrophys.J. 122, 293.
Parker, E.N. 1979, Cosmical magnetic fields, Clarendon Press, Oxford.
Parker, E.N. 1981, Geophys.Astrophys.Fluid Dyn. 18, 175.
Pitts, E. & Tayler, R.J. 1985, Mon.Not.Roy.Astron.Soc. 216, 139.
Priest, E.R. 1985, Rep.Prog.Phys. 48, 955.
Raisbeck, G.M., Yiou, F., Fruneau, M., Loiseaux, J.M., Lieuvin, M., Ravel, J.C. & Lorins, C. 1981, Nature 292, 825.
Rosner, R. & Weiss, N.O. 1985, Nature 317, 790.
Roxburgh, I.W. 1986, in Seismology of the sun and the distant stars, ed. D.O.Gough, p.249, Reidel, Dordrecht.
Ruzmaikin, A.A. 1986, Solar Phys. 100, 125.
Schüssler, M. 1975, Astron.Astrophys. 38, 263.
Schüssler, M. 1981, Astron.Astrophys. 94, 755.
Spörer, G. 1889, Nova Acta Ksl.Leop.-Carol. Deutschen Akad.Naturforscher 53, 272.
Stauffer, J.R., Hartmann, L.W., Burnham, J.N. & Jones, B.F. 1985, Astrophys.J. 289, 247.
Stauffer, J.R., Hartmann, L.W., Soderblom, D.R. & Burnham, J.N. 1984, Astrophys.J. 280, 202.
Stuiver, M. 1980, Nature 286, 868.
Stuiver, M. & Grootes, P.M. 1981, in The Ancient Sun, ed. R.O.Pepin, J.A.Eddy & R.B.Merrill, p.165, Pergamon, New York.
Stuiver, M. & Quay, P.D. 1980, Science 207, 11.
Weiss, N.O. 1985, J.Stat.Phys. 39, 477.
Weiss, N.O., Jones, C.A. & Cattaneo, F. 1984, Geophys.Astrophys.Fluid Dyn. 30, 305.
Weiss, J.E. & Weiss, N.O. 1979, Q.J.Roy.Astron.Soc. 20, 115.
Williams, G.E. 1981, Nature 291, 624.
Williams, G.E. 1985, Aust.J.Phys. 38, 1027.
Williams, G.E. & Sonett, C.P. 1985, Nature 318, 523
Yoshimura, H. 1981, Astrophys.J. 247, 1102.
Zahn, J.-P. 1983, in Astrophysical processes in upper main sequence stars, ed. A.N.Cox, S.Vauclair & J.-P.Zahn, p.253, Swiss Soc. Astron.Astrophys., Geneva.

Modelling of the Magnetic Field of Solar Prominences

U. Anzer

Max-Planck-Institut für Physik und Astrophysik, Institut für Astrophysik, Karl-Schwarzschild-Str. 1, D-8046 Garching, Fed. Rep. of Germany

Measurements of magnetic fields in quiescent prominences have a long tradition. Rust (1967) and Harvey (1969) made extensive studies of strengths and distributions of fields in prominences. Since they used the longitudinal Zeeman effect they could only obtain the line of sight component of the magnetic field $B_{\parallel}$. They found typical field strengths of about 20 Gauss in quiescent prominences. They did not say anything about the direction of the field with respect to the long axis of these prominences. Based on the large number of observational data Anzer and Tandberg-Hanssen (1971) took a statistical approach to find a general trend for this field orientation. They found that in general the field crosses the prominence at a small angle of the order of 15°.

Recently the Hanle effect was incorporated in the field measurements on the solar limb (Sahal-Bréchot et al. 1977; Leroy et al. 1983). The Hanle effect allows one to derive the field component perpendicular to the line of sight, $\underset{\sim}{B}_{\perp}$. Together with the Zeeman effect one can then obtain the total field vector $\underset{\sim}{B}$. Unfortunately the Hanle effect leaves one with the ambiguity that both $\underset{\sim}{B}_{\perp}$ and $-\underset{\sim}{B}_{\perp}$ give the same signal. This ambiguity is a nuisance when one tries to derive the overall structure of the magnetic field configurations associated with quiescent prominences. Photospheric observations taken when these prominences are close to the central meridian show that they always follow a line which divides opposite photospheric polarities. The magnetic field then can be closed either in the way proposed by Kippenhahn and Schlüter (1957) or as Kuperus and Raadu (1974) suggested. The cross sections through both configurations are shown in Fig. 1. Which configuration applies to an individual prominence can in general not be answered because one has the ambiguity between $\underset{\sim}{B}_{\perp}$ and $-\underset{\sim}{B}_{\perp}$ and because prominences are preferentially oriented at small angles with respect to the E-W direction. Leroy et al. (1984) applied a statistical analysis to their large sample of several hundred prominences and found that only about 25% are of the Kippenhahn-Schlüter type, the remaining 75% of the

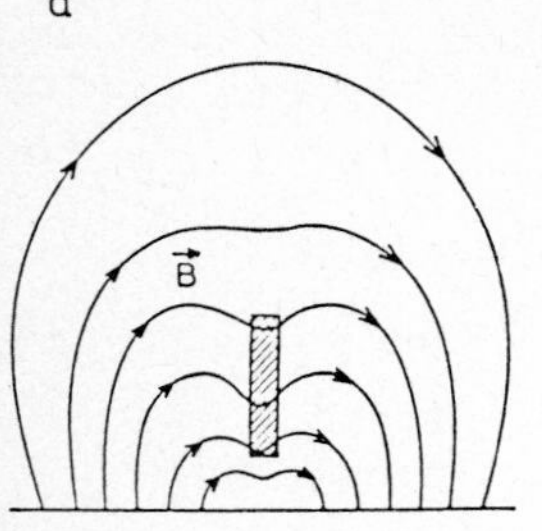

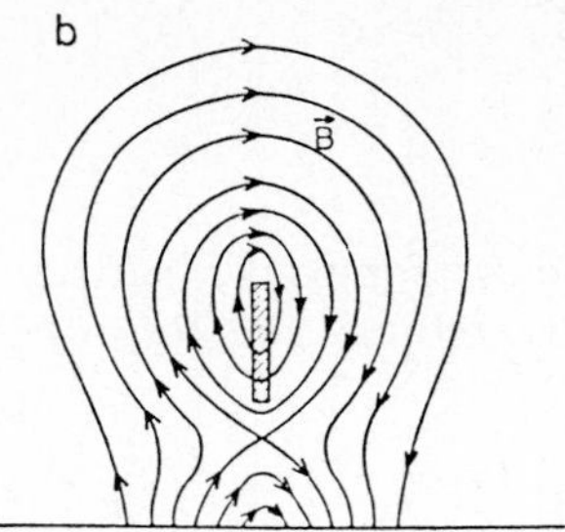

Fig. 1: (a) The Kippenhahn-Schlüter configuration.
(b) The Kuperus-Raadu configuration, adapted from a sketch by Pneuman (1983).

Kuperus-Raadu type. Such statistical analyses are very useful but not entirely compelling because of the assumptions which have to be made. But there is a small subsample of prominences in the observations by Leroy et al. where the prominence lies sufficiently close to the N-S direction such that both $\underset{\sim}{B}_\perp$ and $-\underset{\sim}{B}_\perp$ lead to the same type of configuration. These direct measurments also show a 3:1 dominance of Kuperus-Raadu configurations over the Kippenhahn-Schlüter type. Therefore the evidence is very strong that many prominences on the sun are of the Kuperus-Raadu type. The problem for the theoretician then is to find equilibrium configurations of this type. Simple two dimensional model investigations have been performed by Anzer (1985) and by Anzer & Priest (1985). The global equilibrium problem can be most easily studied by representing the total current flowing in the prominence by a line current I_1. The photospheric field distribution is represented by a subphotospheric line current I_o. In order to obtain the field reversal required by the observations I_1 has to be opposite to the current direction in the Kippenhahn-Schlüter model. This then implies that the resulting Lorentz force is directed downward. This leads to a downward motion of the prominence which will induce currents in the photosphere. These currents have a repulsive effect. If the prominence starts out at a height z_1 and the subphotospheric current is placed at $z = -z_o$ one can estimate at which height $z_1{}^*$ the prominence is in equilibrium. For this the Lorentz force has to be directed upward

$$\underset{\sim}{I} \times \underset{\sim}{B} \,|_z > 0 \tag{1}$$

or

$$-I_1 \, B_x > 0 \ . \tag{2}$$

From

$$B_x = \frac{I_o}{z_1{}^*+z_o} + \frac{I_1}{z_1{}^*+z_1} - \frac{I_1}{2z_1{}^*} \tag{3}$$

one obtains

$$\frac{I_o}{z_1^*+z_o} + \frac{I_1}{z_1^*+z_1} < \frac{I_1}{2z_1^*} . \tag{4}$$

If one takes the following values as representative for the system $I_1/I_o = 1/2$, $z_o = 30000$ km and $z_1 = 240000$ km one finds that $z_1^* \lesssim 9000$ km has to hold. This is by far too low for a quiescent prominence. If one lowers the values of I_1/I_o and z_1 to more realistic ones then z_1^* would have to be even lower. The outcome therefore is that although such equilibria seem possible the heights required are inconsistent with the observations. The more detailed study by Anzer and Priest (1985) supports these results.

The representation of prominences by line currents can only give answers to the questions of the global behaviour of the system, but not about the structure of the prominence itself. Quiescent prominences are known to be thin structures with large vertical extent. Therefore they are more appropriately described by a current sheet. We shall use a very simple sheet extending from $z = a$ to $z = b$ having the following current distribution

$$j = I_1\, 6\, \frac{(b-z)(z-a)}{(b-a)^3} \tag{5}$$

which is normalized to

$$\int_a^b j(z')\, dz' = I_1 .$$

From

$$B_x = - \int_a^b \frac{j(z')(z'-z)}{x^2 + (z'-z)^2}\, dz' \tag{6}$$

one then obtains for $x \to 0$

$$B_x\,(0,z) = -I_1 \frac{6}{b-a} \left[\frac{1}{2}\, \frac{b+a-2z}{b-a} + \frac{(b-z)(z-a)}{(b-a)^2} \ln \left|\frac{b-z}{a-z}\right|\right] . \tag{7}$$

The function in brackets has a minimum of about -0.6 and therefore

$$B_x(0,z)\,|_{max} \approx 3.6\, \frac{I_1}{b-a} \tag{8}$$

holds.

One also finds that far away from the sheet (i.e. for $r > b-a$) the field is very close to that of a line current I_1. Therefore we shall describe the mirror effects of the photosphere by line currents of strength $\pm I_1$, at $z = -z_1$ and $z = -z_1^*$ respectively. The resulting field then is the one given by Eqn. (3), but in addition we now have to include the field of the current sheet. With $a = z_1^*-\delta$ and $b = z_1^*+\delta$ we then obtain

$$B_x(\text{sheet}) = -\frac{3I_1}{\delta}\left[\frac{z_1^*-z}{\delta} + \frac{z_1^*+\delta-z}{2\delta}\,\frac{z-z_1^*+\delta}{2\delta}\ln\left|\frac{z_1^*+\delta-z}{z_1^*-\delta-z}\right|\right]. \quad (10)$$

In this case now the condition for prominence support is that $F_z > 0$ must hold everywhere in the sheet (i.e. for $z_1^*-\delta < z < z_1^*+\delta$) and since the maximum B_x produced by the sheet is $1.8\ I_1/\delta$ one has the condition

$$-\frac{I_o}{z_1^*+z_o} - \frac{I_1}{z_1^*+z_1} + \frac{I_1}{2z_1^*} > \frac{1.8\ I_1}{\delta}\ . \quad (11)$$

From this then one easily obtains a simple necessary (but not sufficient) condition

$$\frac{I_1}{2z_1^*} > \frac{1.8\ I_1}{\delta}\ . \quad (12)$$

Since the whole prominence must lie above the surface of the sun $z_1^* > \delta$ must hold, and therefore condition (12) cannot be fulfilled. Physically this means that in such a configuration the upper parts of the prominence will be pulled downward and the prominence cannot exist as an extended vertical sheet.

Our investigations show that it is not possible to construct simple equilibrium models for quiescent prominences of the Kuperus-Raadu type. An alternative to the closed field structures discussed here are open fields. An example of such a configuration can be found in Malherbe and Priest (1983) and is shown in Fig. 2. The idea is that the field supporting the prominences is not anchored in the solar photosphere but is instead interacting with the solar wind. The force which keeps the prominence in place then has to come from the momentum of the solar wind. Such an equilibrium would therefore have to be a dynamical one. It seems doubtful whether or not such equilibria can exist at all. An

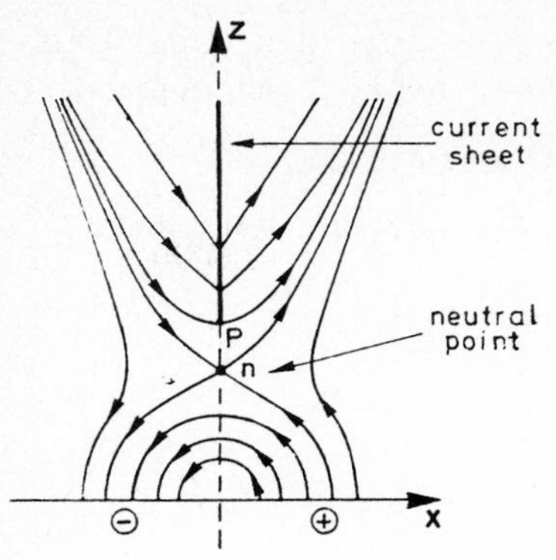

Fig. 2:
Open field configuration, taken from Malherbe and Priest (1983).

additional requirement would be that the equilibrium has to be stable over the life time of prominences (e.g. many months). But most crucially such models are in conflict with another result of the observations by Leroy et al. They found that the prominence field always has a large B_y component (typical angles between prominence axis and field are $\sim 15^o$). The only way how one could produce such a behaviour in these models is by postulating that the solar wind is almost horizontal (instead of vertical) and of opposite direction on the two sides of a prominence. Therefore such models have to be ruled out as well.

One has to conclude from all this that at present there exists no theoretical equilibrium model of the Kuperus-Raadu type which is consistent with the observations. This means that more theoretical work has to be done. But also observationally there are still some open questions. A crucial point is the comparison of the field vector derived from the Hanle effect with the orientation of H_α fibrils near filaments. Zirin (1972) has pointed out that these fibrils line out the magnetic field structure. Therefore such a comparison could give strong support for the existence of Kuperus-Raadu configurations if $\underset{\sim}{B}$ is parallel to the H_α structure. This comparison still needs to be done.

References

Anzer, U.: 1985, Proceedings of the MSFC Workshop on Measurements of Solar Vector Magnetic Fields, 101

Anzer, U., Priest, E.: 1985, Solar Phys. 95, 263

Anzer, U., Tandberg-Hanssen, E.: 1971, in IAU Symposium No. 43, Solar Magnetic Fields, Ed. R. Howard, 656

Harvey, J.W.: 1969, Ph.D. Thesis, University of Colorado

Kippenhahn, R., Schlüter, A.: 1957, Z. Astrophys. 43, 36

Kuperus, M., Raadu, M.A.: 1974, Astron. Astrophys. 31, 189

Leroy, J.L., Bommier, V., Sahal-Bréchot, S.: 1983, Solar Phys. 83, 135
Leroy, J.L., Bommier, V., Sahal-Bréchot, S.: 1984, Astron. Astrophys. 131, 33
Malherbe, J.M., Priest, E.R.: 1983, Astron. Astrophys. 123, 80
Rust, D.M.: 1967, Astrophys. J. 150, 313
Sahal-Bréchot, S., Bommier, V., Leroy, J.L.: 1977, Astron. Astrophys. 59, 223.
Zirin, H.: 1972, Solar Phys. 22, 34.

On the Frequencies of Solar Oscillations

M. Stix[1] *and M. Knölker*[2]

[1]Kiepenheuer-Institut für Sonnenphysik, Schöneckstr. 6, D-7800 Freiburg i. Br., Fed. Rep. of Germany

[2]Universitäts-Sternwarte, Geismarlandstr. 11, D-3400 Göttingen, Fed. Rep. of Germany

Solar Oscillations, with frequencies between 2 and 5 mHz, can be identified as p modes with well-determined degree l and overtone number n, but minor discrepancies, of order 10 μHz, between observed and calculated frequencies remain. We describe the computation of solar models and their frequencies of oscillation, check the accuracy of the numerical results, and study the influence of the atmosphere. Attempts to improve the calculated frequencies for low degree and intermediate order (n=10 ...20) have so far been unsuccessful.

1. INTRODUCTION

Oscillatory motions in the solar atmosphere were discovered in 1960 by R.B. Leighton who used photographic differences of spectroheliograms obtained in the red and blue wings of spectrum lines (the "Doppler-plates", cf. Leighton et al., 1962). But only in 1975 it was confirmed by F.L. Deubner that these oscillations, which previously had been represented by a broad frequency distribution (periods between, roughly, 3 and 10 minutes) are in fact a superposition of a large number of eigenmodes with discrete frequencies, as predicted by Ulrich (1970). Since then, many such "p mode" frequencies have been measured. For the present contribution the case of low degree, l, is of particular interest (l is the total number of node circles of a particular eigenoscillation on the solar surface). Although single l-values are not directly resolved by means of the spherical harmonic analysis, single modes can be identified with the help of the resolution in frequency. Thus, tables of frequencies have been presented, and the frequency resolution reached by the diverse observations, and also the comparison of results of various observers (e.g. Grec et al., 1983; Duvall and Harvey, 1983; Harvey and Duvall, 1984; Libbrecht and Zirin, 1986) ensures that the accuracy of the known low l eigenfrequencies is about 1 μHz or better.

Theoretically, the eigenfrequencies are obtained from an adiabatic perturbation of the equations of stellar structure. Reflecting layers, in particular the decrease of the velocity of sound towards the solar surface, provide the boundaries required for a discrete spectrum. The problem is solved by asymptotic (WKB) or numeric methods, and the frequencies so obtained more or less coincide with the observed ones so that an unambiguous identification of p mode degree, l, and order (overtone number), n, is possible. Minor differences do however remain. Here we will be concerned in particular with the discrepancy of up to about 10 μHz for p modes with n=12 ... 25. This discrepancy is "significant and unresolved", as pointed out by Noyes and Rhodes (1984); since it is similar for all small l values we shall restrict the subsequent discussion to the case l=0, i.e. the radial pulsations, in which the mathematical treatment is more convenient.

2. SOLAR MODELS

Before the oscillations can be computed one must have an equilibrium model of the Sun. We have computed such models with the program written by Kippenhahn, Hofmeister and Weigert (1967). The present workshop certainly is an appropriate occasion to thank these authors for their generous way to let us and other groups benefit from their work. The model calculation is done by the usual procedure of evolving an initially homogeneous star, of 1 solar mass and helium content Y_o, for a period equal to the Sun's age, for which we adopted $4.64 \cdot 10^9$ years. The value of Y_o is then, together with the ratio, α, of mixing length to scale height, calibrated so that the most recent model of the evolutionary sequence has the observed luminosity and radius. The calibration yields $Y_o \approx 0.28$, and $\alpha = 1.3 \ldots 1.5$, depending on what we do with the equation of state and several other model ingredients (see below). Our original helium content is rather close to the one obtained by Noels et al. (1984), but somewhat larger than the $Y_o \approx 0.25$ of Christensen-Dalsgaard (1982) and Ulrich (1982) and still larger than the $Y_o = 0.236$ of Demarque et al. (1986). Possible sources of such differences (R.K. Ulrich, 1986, private communication) are a too course temperature-density grid in the opacity tables and/or the neglect of a correction $-0.07(1+X)$ to the opacity, which Bahcall et al. (1982) suggested in order to account for collective effects on electron scattering. As a test we have calculated a solar model including the latter correction and obtained $Y_o = .272$, and a reduction of the neutrino rate by ca. 10%. However, the p mode eigenfrequencies change very little, by 1 μHz approximately; the conclusions below remain therefore unaffected. Incidentally it seems that the larger value of Y_o is quite consistent with estimates of the galactic helium abundance based on radio astronomical results (Thum et al., 1980).

The opacity in our models is interpolated between 3 Los Alamos tables (Huebner et al, 1977) of different helium content and a table of Kurucz (1979) for the cool outer layers (cf. Knölker and Stix, 1984; we thank the authors for kindly communicating their material to us). For the abundance of elements heavier than helium we use Z = .017. Ionization of HI, HeI and HeII is treated explicitely by means of Saha equations, but the partition functions were simply replaced by the weight of the respective ground states. Full ionization was assumed for the inner parts of the model, as well as everywhere for all the elements represented by Z. All models are standard models, also in the sense that they predict the usual high neutrino rates: we obtain 8 snu for the ^{37}Cl experiment.

3. OSCILLATIONS

The oscillation equations have been written down frequently, our version e.g. by Knölker (1983) or Knölker and Stix (1983). Here we shall therefore be content with a symbolic form. We shall, however, present two such forms; the usefulness of this will be clear presently. The first method, which we call the "Henyey Method", is obtained by solving the perturbation equations (for each l, i.e. after separation of the horizontal dependence) for the <u>depth</u> derivatives of $x = \xi_r/r$ and $p = \delta P/P$ (ξ_r is the radial displacement, P the pressure, and δP its Lagrangian perturbation). Let a prime denote d/dlnP, then

$$\begin{pmatrix} x' \\ p' \end{pmatrix} = L_1(\sigma^2)\begin{pmatrix} x \\ p \end{pmatrix} \tag{1}$$

L_1 is linear operator on x and p, but depends on σ^2, the square of a dimensionless frequency, in a non-linear fashion. For $l \neq 0$ the system (1) is replaced by a system for four variables, but can again be reduced to the simple form (1) through the neglect of the Eulerian perturbation of the gravitational potential. This approach, called the "Cowling Approximation" (cf. Cowling, 1941), gives very accurate results when l is large (Christensen-Dalsgaard, 1984). - Here we stay with l=0, where (1) is exact. The procedure then is to guess an eigenvalue σ^2, and eigenfunctions x and p, an to iterate. For this a generally applicable Henyey routine, named GNR1 and written by L. Lucy and N. Baker, worked quite well. As the authors are here, we express our thanks to them for letting their program circulate so generously among astronomers. Usually 3 to 5 iterations gave frequencies with relative errors below 10^{-5}, and only very rarely a high overtone was not found from the first guess.

In the second method, referred to as the "matrix method", the perturbation equations are solved for the <u>time</u> derivatives:

$$\sigma^2 \begin{pmatrix} u \\ v \end{pmatrix} = L_2 \begin{pmatrix} u \\ v \end{pmatrix} \tag{2}$$

The variables u and v are linear combinations of x and p, chosen so that the linear operator L_2 - after introduction of finite differences for the depth derivatives - becomes a symmetric matrix. This is always possible because L_2 is self-adjoint for adiabatic equations and energy-conserving boundary conditions. For $l=0$, or for $l \neq 0$ in the Cowling Approximation, the matrix is tridiagonal. No first guesses of eigenvalues or eigenfunctions are necessary, and EISPACK routines, using bisection, rapidly yield all eigenvalues in any desired interval. In fact we have combined our two methods: The frequencies found by the matrix method were used as first guesses for the Henyey method.

The typical result of an eigenvalue search is shown in Fig. 1. The presentation is in form of the "Echelle diagram", where the whole spectrum is cut into pieces of equal length, here 136 μHz, which then are arranged in lines on top of each other. This is very illustrative for spectra with (almost) equidistant lines, such as the solar p modes.

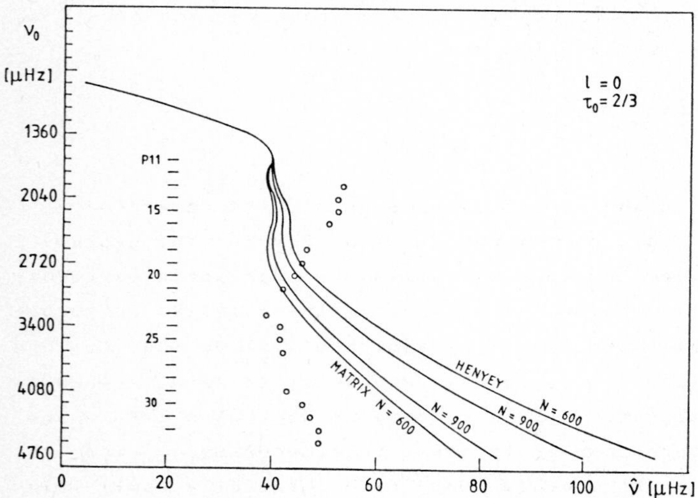

<u>Figure 1.</u> Echelle diagram for radial p modes. The lines connect the calculated frequencies, $\nu = \nu_o + \hat{\nu}$, i.e. abscissa + ordinate . The outer boundary condition is placed at optical depth $\tau_o = 2/3$; N is the number of grid points in depth. The open circles are the frequencies observed by Grec et al. (1983).

Three points are immediately visible: first, the freqencies of the modes lower than, say, p_{20} are <u>smaller</u> than their observed counterparts; second, overtones higher than p_{20} have <u>larger</u> than observed frequencies; and third, with increasing mode number there is an increasing dependence on the total number, N , of grid points used in the finite difference schemes for (1) and (2).

Let us discuss the last point first. As pointed out by Shibahashi and Osaki (1981), an extrapolation to $N=\infty$ must be made in order to avoid significant errors. Figure 2 demonstrates that such an extrapolation should be safe. As the errors of the simple centered differences used in both our methods are proportional to h^2, where $h= \Delta \ln P \sim 1/N$ is the step width, we expect $\nu(N) - \nu(\infty) \sim h^2$, and this is indeed the case. Moreover, the frequencies obtained from the two methods intersect the line h=0 within, say, 1 μHz.

Now the second point. The higher overtones of Fig. 1 have too large frequencies because the atmosphere is not properly included in this example, in which the outer boundary condition, $\delta P=0$, was imposed at optical depth $\tau_o=2/3$. For smaller values of τ_o there is better agreement with the observations, cf. Fig. 3. The reason is that the time of sound propagation across the Sun, i.e. the integral occuring in the frequency splitting

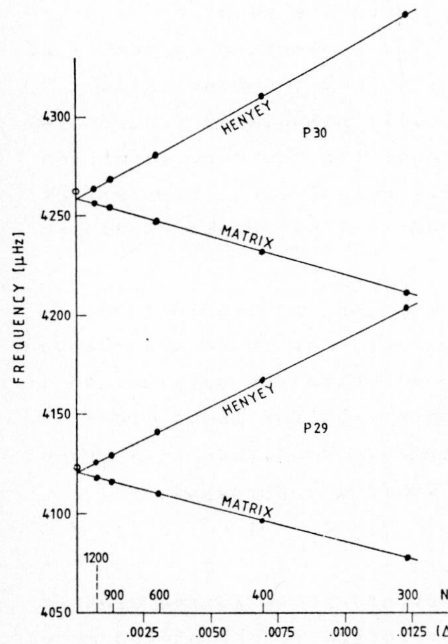

<u>Figure 2.</u> Frequencies of radial p modes of order 29 and 30, as functions of the number, N, of grid points in depth. The outer boundary is placed at $\tau_o=10^{-4}$. The open circles mark the observed frequencies.

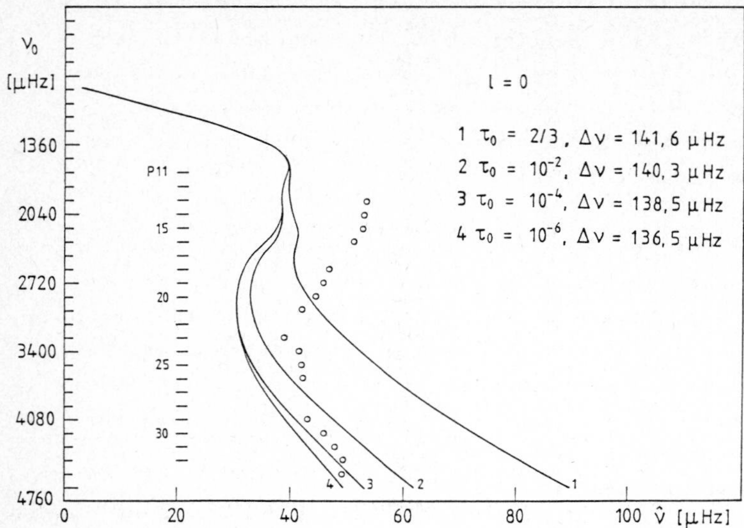

Figure 3. Echelle diagram for radial p modes, as Fig. 1. The frequencies are extrapolated to $N \to \infty$; τ_o is the optical depth where the boundary condition is applied, and $\Delta\nu$ is the asymptotic frequency spacing, calculated according to (3).

$$\Delta\nu = \left(2\int_0^{r_\odot} \frac{dr}{c} \right)^{-1} \qquad (3)$$

gets its largest contribution in the layer where the velocity of sound, c, is smallest, that is in the atmosphere. The respective values of $\Delta\nu$ given in Fig. 3 are obtained via (3) directly from the equilibrium model and confirm the splitting of the actually calculated frequencies. For the subsequent calculations we have placed the boundary condition at $\tau_o = 10^{-4}$. Still smaller choices of τ_o have relatively little effect on the frequencies (cf. also Ulrich and Rhodes, 1983; Noels et al, 1984).

The lower p modes do not vary as τ_o is changed. An argument of Christensen-Dalsgaard and Gough (1980) says why: The lower the frequency, the deeper in the atmosphere is the level where the oscillation is totally reflected (becomes evanescent). Hence, for the lower modes it does not matter where exactly the outer boundary condition is imposed, if only the exponentially growing part is somehow suppressed.

4. THE EQUATION OF STATE

We must now discuss the first point mentioned above in the context of Fig. 1: the too small frequencies of theoretically calculated modes below $\approx p_{20}$. This is the discrepancy described in the introduction; it is confirmed by the present calculations and seems to be quite persistent. The only calculated frequencies which do better agree with the observations are those of Shibahashi et al. (1983). It is, however, unclear to which of several simultaneous changes (of the earlier model of Shibahashi and Osaki, 1981) the improvement must be attributed: the introduction of Los Alamos opacity tables (Huebner et al., 1977), the electrostatic correction in the equation of state, or the calculation of partition functions in the Planck-Larkin approximation. The latter modification has not yet been included in our own calculations. Ulrich and Rhodes (1984a) find, however, that its effect on the eigenfrequencies in question is much smaller than the derived improvement. - The opacity in our own model is also from the Los Alamos Library. The electrostatic correction, then, should be the main reason for the shifted frequencies of Shibahashi et al. (1983). Fig. 4 demonstrates however that this effect alone is again unable to bring theory and observation into agreement. The improvement is 3 μHz at most, and the results of Shibahashi et al. remain unclear. Their models have, in addition, rather large central temperatures ($>1.6 \cdot 10^7$K) and central densities (>185 gcm^{-3}); we suppose that the neutrino production, had it been calculated, would be still larger than the already large 6 to 8 snu s of the usual standard models.

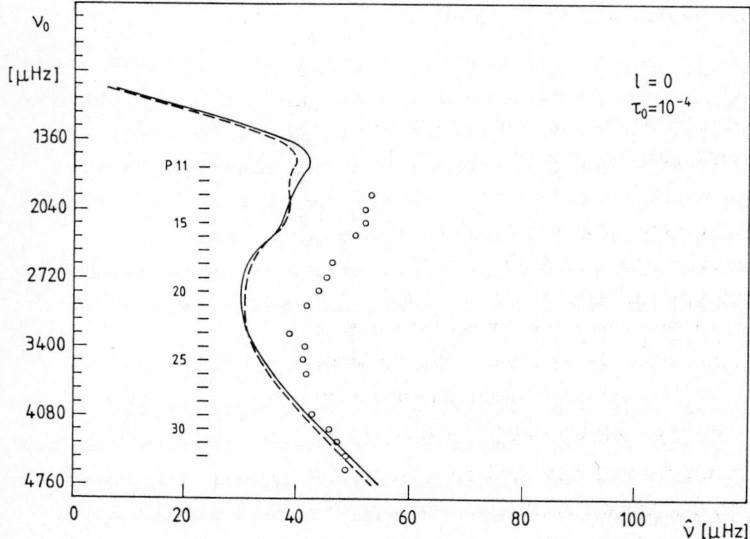

Figure 4. Echelle diagram for radial p modes, as Fig. 1. The solid (dashed) line connects the extrapolated frequencies calculated with (without) the electrostatic correction.

With the electrostatic correction included according to the Debye-Hückel theory, the gas pressure is

$$\beta P = \frac{\varrho \mathcal{R} T}{\mu} - \frac{kT}{24\pi r_D^3} \tag{4}$$

where $r_D(T,\varrho)$ is the Debye radius, and β (≈ 1 for the Sun) is the ratio of gas pressure to total pressure. The correction is negative, in accord with the possibility to sink energy into the charged clouds of size r_D^3. As it turns out the correction is largest at the depth where $T \sim 5 \cdot 10^4$ K and is roughly -7% in the models of Shibahashi et al., and -5.5% in our own calculations. It is therefore plausible that it affects the lower p modes which are insensitive to manipulations of the atmosphere. But one may ask why the effect is to <u>increase</u> the frequency when the pressure (the "elasticity" of the gas) suffers a <u>negative</u> correction. The answer is that in the stellar model the total pressure is determined by the hydrostatic equilibrium, and hence should be the same wether or not the correction is applied. This is confirmed by the numerical calculations: the electrostatic correction is indeed compensated by a corresponding increase of the perfect gas term in (4). At the depth where the effect is largest, for example, this is accomplished through a change of ϱ by 1.7%, of T by 2.6%, and of μ by -1.1%. With the temperature the speed of sound is also slightly increased, resulting in a somewhat higher eigenfrequency of the oscillations.

We have made a number of other numerical experiments. But the eigenfrequencies rather firmly stay where they are in Fig. 1: A 10% opacity decrease has virtually no influence (although it lowers the neutrino rate from 8 to 6 snu); when the assumption that the elements heavier than He are fully ionized in the Sun´s interior is replaced by the other extreme of no ionization, the effect is equally small; and a simple power law for the energy generation (instead of the more detailed temperature dependence of the nuclear reaction rates) raises the frequencies by only 3 μHz or less.

Of course, it is good that the eigenfrequencies are so stable against changes of the solar model, because this means that we can possibly learn something from the discrepancy between theory and observation. Perhaps it is necessary to construct solar models which have deeper convection zones (the one of Shibahashi et al. has!). But the depth of the convection zone by itself is not a free parameter of the model. Ulrich and Rhodes (1984b) have proposed that a non-local version of the mixing length formalism could help. This would generate an overshoot layer at the base of the convection zone. Pidatella and Stix (1986) made such a calculation with an envelope program, and we plan to incorporate this into the stellar evolution code.

Perhaps more exotic steps are necessary. WIMPs - weakly interacting massive particles - have recently been proposed as a remedy to the solar neutrino discrepancy, and WIMP models also very accurately account for the observed frequency splitting between $p_{n,l}$ and $p_{n-1,l+2}$ modes (Faulkner et al., 1986; Däppen et al., 1986). We wonder wether the discrepancy discussed in the present contribution also would be removed in such models.

REFERENCES

Bahcall, J.N., Huebner, W.F., Lubow, S.H., Parker, P.D., Ulrich, R.K.: 1982, Rev. Mod. Phys. 54, 767

Christensen-Dalsgaard, J.: 1982, Mon. Not. R. astr. Soc. 199, 735

Christensen-Dalsgaard, J.: 1984, 25th LIEGE Coll., p 155

Christensen-Dalsgaard, J., Gough, D.O.: 1980, Nature 288, 544

Cowling, T.G.: 1941, Mon. Not. R. astr. Soc. 101, 367

Däppen, W., Gilliland, R.L., Christensen-Dalsgaard, J.: 1986, Nature 321, 229

Demarque, P., Guenther, D.B., van Altena, W.F.: 1986, Astrophys. J. 300, 773

Deubner, F.-L.: 1975, Astron. Astrophys. 44, 371

Duvall, T.L., Harvey, J.W.: 1983, Nature 302, 24

Faulkner, J., Gough, D.O., Vahia, M.N.: 1986, Nature 321, 226

Grec, G., Fossat, E., Pomerantz, M.: 1983, Sol. Phys. 82, 55

Harvey, J.W., Duvall, T.L.: 1984, Snowmass Conf. Solar Seismology from Space, R.K. Ulrich (ed.), JPL, p. 165

Huebner, W.F., Merts, A.L., Magee, N.H., Jr., Argo, M.F.: 1977, Astrophysical Opacity Library, Los Alamos Scientific Laboratory report LA-6760

Kippenhahn, R., Weigert, A, Hofmeister, E.: 1967, Meth. Comp. Phys. 7, 129

Knölker, M.: 1983, Dissertation, Univ. Freiburg

Knölker, M., Stix, M.: 1983, Solar Phys. 82, 331

Knölker, M., Stix, M.: 1984, Mem. Soc. Astr. Ital. 55, 305

Kurucz, R.L.: 1979, Astrophys. J. Suppl. Ser. 40, 1

Leighton, R.B., Noyes, R.W., Simon, G.W.: 1962, Astrophys. J. 135, 474

Libbrecht, K.G., Zirin, H.: 1986, Astrophys. J. in press

Noels, A., Scuflaire, R., Gabriel, M.: 1984, Astron. Astrophys. 130, 389

Noyes, R.W., Rhodes, E.J.: 1984, Probing the Depths of a Star: The Study of Solar Oscillations from Space, NASA-report

Pidatella, R.M., Stix, M.: 1986, Astron. Astrophys. 157, 338

Shibahashi, H., Osaki, Y.: 1981, Publ. Astron. Soc. Japan 33, 713

Shibahashi, H., Noels, A., Gabriel, M.: 1983, Astron. Astrophys. 123, 283

Ulrich, R.K.: 1970, Astrophys. J. 162, 993

Ulrich, R.K.: 1982, Astrophys. J. 258, 404

Ulrich, R.K., Rhodes, E.J.: 1983, Astrophys. J. 265, 551

Ulrich, R.K., Rhodes, E.J.: 1984a, 25th LIEGE Coll, p. 250

Ulrich, R.K., Rhodes, E.J.: 1984b, Snowmass Conf. Solar Seismology from Space, R.K. Ulrich (ed.), p. 371

Thum, C., Mezger, P.G., Pankonin, V.: 1980, Astron. Astrophys. 87, 269

NOTE ADDED AFTER COMPLETION OF THIS WORK

In the models discussed in this paper the electrostatic pressure correction was not applied in the central part ($r \lesssim 0.5\, r_\odot$). Although the correction at this depth is small ($\approx 1\%$), it requires a noticable adjustment of the helium content ($Y_0 \approx 0.26$). The neutrino rate comes down to 6.6 snu, and the sound velocity is raised by 0.4% in the average. This is turn raises the eigenfrequencies. Of all our models calculated so far, this one has the smallest discrepancy to the observed l=0 frequencies of oscillation.

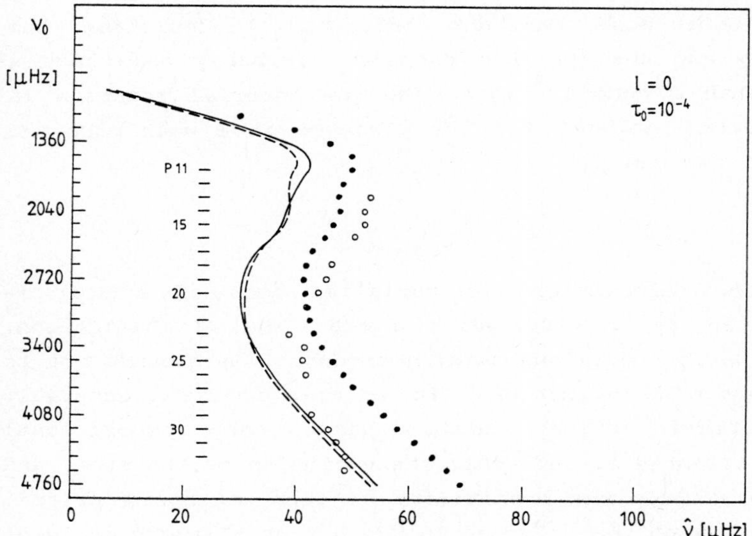

Figure 5. Echelle diagram for radial p modes, as Fig. 4. The dots are the frequencies of a model which has the electrostatic correction incorporated throughout.

Mixing and Transport of Angular Momentum in the Solar Interior

H.C. Spruit

Max-Planck-Institut für Physik und Astrophysik, Institut für Astrophysik, Karl-Schwarzschild-Str. 1, D-8046 Garching, Fed. Rep. of Germany

The observational indicators of mixing in the solar radiative interior, and the known mechanisms for mixing are reviewed critically. It is concluded that the three most important indicators, Li-depletion, the neutrino discrepancy and the internal rotation, probably each have a different, rather than a common origin. The low internal rotation is interpreted as a strong evidence for the presence of a weak magnetic field in the core of the sun.

1. INTRODUCTION.

In stellar evolution calculations, a radiative zone of a star is traditionally treated as a very quiet place. The stratification, especially that of nuclear fuel and burning products, is assumed not to be disturbed by random fluid motions. It is also realized generally that very weak fluid motions could change the compositional stratification sufficiently to influence the evolution of the star, and that it is hard to exclude the possibility of such motions a priori. For example, a velocity on the order of 10^{-6}cm s^{-1} on a length scale of 10^5 km would be enough to keep the interior of the sun mixed, thereby preventing it from becoming a red giant.

Interest in the effects of a moderate amount of mixing in stellar interiors has been stimulated by a series of papers by Schatzman and his coworkers (Schatzman, 1977, Schatzman and Maeder, 1981, Genova and Schatzman, 1979, Bienaymé et al. 1983). Without going into the details of how the mixing takes place, they assumed that differential rotation leads to turbulence with a turbulent diffusivity ν_t=Re*ν, where ν is the ordinary (kinematic) viscosity of the plasma, and Re* weakly varying number. By taking Re*$\sim$100, an impressing number of known discrepancies could be solved, ranging from the observed width of the upper main sequence to the low value of the solar neutrino flux. In view of these successes, it seems appropriate to study in detail which mixing processes can occur in a star, and why the effective diffusivity resulting from these processes should be a weakly varying multiply of the kinematic viscosity. In this paper I review the extent to which these questions can be answered at present.

2. EVIDENCE FOR MIXING.

The most direct evidence for mixing in the radiative interior of the sun is the observed Li-depletion. The temperature of about 2.5 10^6K at which Li burns is not reached within the convection zone. The most stringent constraints are placed by the observed dependence of [Li] on spectral type along the main sequence (Cayrel et al. 1984 and references therein, Duncan and Jones, 1983, Soderblom, 1983, Boesgaard and Tripicco, 1986; Michaud, 1986). This dependence is smooth, in contrast with the expected very sharp jump in depletion at the spectral type for which the convection zone just reaches 2.5 10^6K. Mixing by convective overshoot is not likely to reproduce the dependence on spectral type satisfactorily (Cayrel et al. 1984). What is needed instead is a low level of mixing which has a relatively uniform diffusivity over a depth range of about 1 scale height below the convection zone (Vauclair et al. 1978, Cayrel et al. 1984). The dependence of the Lithium depletion on age (Skumanich, 1972, Duncan, 1981, Soderblom, 1983) also provides very interesting information on the mixing mechanism. Since this dependence is also rather smooth but not exponential (perhaps following Skumanich's $t^{-1/2}$ law), the diffusion coefficient cannot be constant in time, but it cannot vary too steeply either.

The Lithium depletion, together with less well-determined observables like the Be depletion and the ^{3}He/^{4}He isotope ratio at the solar surface (Schatzman and Maeder, 1981), only provide evidence for mixing in the outermost layers of the radiative core. In deeper layers (T>3.5 10^6K, the observed rotation rate (Duvall and Harvey, 1984, Duvall et al. 1984, Brown, 1985) and the solar neutrino flux provide information which may in principle be explained by some mixing though other plausible processes exist as well.

The very low rotation rates deduced by Duvall and Harvey (almost uniform except for a possible increase below $r/R_{\odot}=0.3$), indicate the existence of and efficient process transporting the angular momentum of the core to the surface during the spindownn of the sun.

A moderate amount of mixing is known to be very effective at reducing the neutrino flux, by the following well known argument (For a review of the neutrino problem see Bahcall et al. 1982). The rate of energy production in the core is proportional to the square of the hydrogen abundance X. If we compare a model in which fresh hydrogen has been mixed down into the core with an unmixed model with the same luminosity (namely, the observed luminosity), the mixed model must have a somewhat lower central temperature. (In constructing models this is done by adjusting the primordial helium abundance slightly.) Since the neutrino flux as detected in the ^{37}Cl experiment is very sensitive to the central temperature, this has a large effect on the neutrino flux. To decrease the flux by a factor of 3, the central hydrogen abundance has to be increased from 0.36 in unmixed models to about 0.62. The required mixing effect is therefore substantial.

3. MECHANISMS FOR MIXING.

The following five mechanisms are potentially available for mixing:

1. Diffusion.
2. Meridional circulation (Eddington-Sweet-Vogt circulation).
3. Turbulence induced by hydrodynamical instability of some large-scale flow (most notably differential rotation).
4. Mixing induced by magnetic forces.
5. Mixing by unstable nuclear burning.

We briefly discuss possibilities 1, 2, 4, 5 before proceeding to the most widely discussed mechanisms in class 3.

3.1. Diffusion.

In the absence of radiative and gravitational forces, diffusion would reduce inhomogeneities in composition, in the presence of gravity, temperature gradients and radiation pressure however, unmixing occurs, because of gravitational settling, differences in the radiative force on different species, and the thermal diffusion effect. In the interiors of main sequence stars, the time scales for diffusive separation are long compared with the main sequence life time. In the atmospheres of stars, the effects can be quite strong, especially in white dwarfs and their precursors. This is because in these cases the relevant length scale, namely the pressure scale height H, is small, so that the diffusion time scale H^2/D (where D is the kinematic diffusivity) can be short. For a review, see Vauclair and Vauclair, 1982. For the problems discussed here, diffusion can be neglected.

3.2. Meridional circulation.

Due to its oblateness, a rotating star is in general not in thermal balance (i.e. $\mathrm{div} F_r \neq 0$). A steady state is maintained instead by a meridional circulation, the Eddington-Sweet or Eddington-Vogt circulation (see, e.g., Zahn, 1983). This steady state evolves secularly (in this case: slowly compared to the thermal time scale) because it transports angular momentum, changing the state of internal rotation, and thereby changing the circulation. After enough time has elapsed, the star may conceivably settle into a new steady state, for example one in which the rotation rate varies in exactly such a way throughout the star that the induced circulation is zero. Since the time required for the circulation to turn over once is in general long compared with the main sequence life of the star (except perhaps in some rapidly rotating A stars), the approach towards such steady solutions is of little relevance and the circulation can in general be calculated without taking the effect of the circulation on the internal rotation field into account. In a problem studied by Busse (1981), a very rapid approach towards a new steady solution without circulations was found, and Busse claimed that Eddington-Sweet-Vogt circulations do not exist. As pointed out by Osaki (1982) however, Busse's problem is not relevant to the astrophysical situation, and his claim is unfounded.

The presence of a stabilizing composition gradient is known to exert a strong influence on the Eddington-Sweet-Vogt circulation. Mestel (1953) showed that a μ-gradient can break up the circulation, and prevent mixing in the core of an evolved star. He also showed that the circulation is not fast enough to keep an initially homogeneous star mixed during its nuclear evolution if the ratio of centrifugal to gravitational acceleration in the core is less than 1/30. These conclusions were challenged by Tassoul and Tassoul (1983), on the basis of one of their own calculations, which did not show the existence critical μ-gradient above which the circulation would be prevented. As argued by Mestel (1984b) however, the Tassouls' calculation is of no direct relevance to the inhibiting effect of a μ-gradient on the circulation, and a μ-gradient is indeed effective in preventing a circulation. The degree to which a μ-barrier can be penetrated by an ESV-circulation was investigated in detail by Huppert and Spiegel (1977). Penetration is prevented, roughly, when

$$\Omega < N_\mu \quad (1)$$

where Ω the rotation rate and N_μ the buoyancy frequency due to the compositional stratification, is given by $N_\mu^2=-g\partial_r \ln\mu$. In a star like the sun, condition (1) is amply satisfied throughout the greater part (by mass) of the core.

All extensive series of calculations of the ESV circulation, the most mathematically consistent so far, has been undertaken by Tassoul and Tassoul (1983 and references therein). Their method consists of an expansion in a small parameter, the ratio of centrifugal acceleration to gravity. Though these solutions are in many ways better than previous results, Zahn (1983) and Roxburgh (1984) point out that they still contain a significant inconsistency, because the diffusion term in the heat transport equation was not treated consistently with that in the equation of motion.

Regarding the mixing effect of ESV circulations, we may conclude firstly that in regions of significant compositional inhomogeneity ($N_\mu > \Omega$), circulations may be safely ignored. Secondly, in slowly rotating stars like the sun, the mixing effect in regions of homogeneous composition is also negligible. As pointed out by Cayrel et al. (1984) and Vauclair et al. (1978), this agrees well with observations of the sun, which show a large Li-depletion, but only a weak Be-depletion. If the ESV circulation were responsible for a significant amount of mixing, the depletions of Be and Li should have been comparable.

A circulation can also be generated by Ekman-pumping, a viscous effect generated by the presence of a jump in the rotation rate. For a discussion see Zahn (1983), its speed can be shown to be at most of the same order as that of the ESV circulation, so that the previous conclusions apply to this effect as well.

3.3. Mixing and spin-down caused by magnetic forces.

Unstable magnetic fields might be able to produce fluid motions, and turbulence, in areas where ordinary fluid instabilities are absent. To guarantee continued mixing, such as required to lower the neutrino flux or to deplete Lithium, the magnetic fields would have to be recreated continually. One source of magnetic flux would be Biermann's battery process (Biermann, 1950, Mestel and Moss 1983), but the amount produced this way is not very large. Another process is winding up of fields by differential rotation. In this case, the energy source for the mixing would be the differential rotation, just like in the case of the hydrodynamic instabilities discussed in section 4. The magnetic field would only play a mediating role.

More important than its mixing effect would be the torques exerted by a magnetic field. Define the "torque-averaged" magnetic field as

$B=\langle B_\phi B_r \rangle^{1/2}$, where B_ϕ is the azimuthal (east-west) component of the field, B_r the radial component, and the average is taken over an equipotential surface in the core. A value of $B \sim 1G$ or larger would be sufficient to transmit the torque exerted by the solar wind to the core, so that the interior of the sun would corotate with the surface during its spindown. This seems a rather modest field strength, and it would seem plausible that the state of internal rotation of the sun is determined mostly by a magnetic field in its interior. On the other hand, very little is certain about the nature of such a field. For a review of current ideas see Mestel (1984a). A scenario for the effect of magnetic fields in the solar interior was proposed by Rosner and Weiss (1985, see also Weiss, this volume).

3.4. Mixing by unstable nuclear burning.

The distribution of ^3He in an evolving solar model is potentially unstable (Dilke and Gough, 1972). Gravity-modes can feed on the energy released in the burning of ^3He, and Dilke and Gough speculated that such modes would slowly grow to large amplitudes followed by an episode of rapid mixing. The process would repeat in the manner of a relaxation oscillation, on a time scale of 10^8yrs. It does not seem necessary, however, that the instability should grow in this way. Alternatively the amplitude of the unstable g-mode could be limited, by nonlinear effects, to a moderate constant value, such that the amount of ^3He burnt due to the oscillation balances the amount produced by H-burning. In this case the mixing effect would be much smaller. Also, it is not sure whether any g-modes can really become unstable, because of uncertainties in the damping of g-modes by the convection zone (Christensen-Dalsgaard et al. 1974).

3.5. Mixing in the core: an energy argument.

Helium nuclei are produced near the center of the sun at the bottom of the gravitational potential well. If the core is mixed, helium nuclei have to be moved upwards in the potential well. Thus mixing of the core requires input of energy, by an amount equal to the difference in binding energy between the mixed and the unmixed models. The amount, for the present sun, can be estimated by noting that an amount of helium of the order of $0.1M_\odot$ has to be displaced over a significant fraction, 0.1 say, of the depth of the potential well. The energy E_m required for mixing is thus of the order.

$$E_m = 10^{-2}E_g \sim 10^{46} \text{ erg}$$

where E_g is the binding energy of the sun. This can be compared, for example, with the present rotational energy content of the sun, E_{rot} = 3 10^{42}erg. If the mixing takes place gradually over the life of the sun, an energy input rate W_m is needed,

$$W_m = E_m/(4\ 10^9 yr) \sim 10^{29}\ erg\ s^{-1},$$

or

$$W_m \sim 2\ 10^{-5}\ L_\odot\ .$$

These amounts of energy are not small, and pose restrictions on potential mixing mechanisms which are proposed for compositionally inhomogeneous parts of the sun. Mixing outside the inhomogeneous core, for example in the Li-depletion zone, is not restricted by this argument.

4. MIXING BY HYDRODYNAMIC INSTABILITIES.

A large scale steady flow or oscillation which itself does not mix anything in a stellar interior may become unstable to a number of different instabilities that could lead to turbulence and mixing.

4.1. Mixing by gravity waves.

The possiblity that internal gravity waves, emitted by the convection zone into the core, would reach amplitudes sufficient for breakdown into turbulence has been considered by Press (1981). He showed that such waves would be focussed towards the center of the sun, and by appealing to the fact that gravity waves break down nonlinearly already at low amplitudes, argued that this could cause significant mixing in the core. Thus, this mechanism would be especially relevant for the neutrino problem. The mechanism should, however, be energetic enough to supply the difference in binding energy between unmixed and partially mixed solar models, as discussed in section 3.5 above. The energy flux in gravity waves can be estimated as follows. The deformation X of the interface between the convection zone due to pressure fluctuations in the convection zone is of the order of X = 200km, and varies on a time scale of the order τ=3 10^6sec. Assuming that all of this interface motion takes the form of freely running gravity waves, their vertical velocity amplitude is $v_r=X/\tau$=6cm s^{-1}, corresponding to an energy flux F of (cf. Press 1981)

$$F = \rho(N^2 - \omega^2)^{½} v_r{}^2/k_H \sim 3\ 10^7\ erg\ cm^{-2}s^{-1}.$$

where $N \sim 10^{-3}$ is a typical value for the buoyancy frequency in the core, $\rho \sim 0.2$ the density at the base of the convection zone, $\omega=2\pi/\tau$, and $k_H \sim 3$

10^{-10} a typical value for the horizontal wavenumber. The total energy input W_w by waves would be $W_w=4\pi r^2 F \sim 10^{30}$erg s^{-1}. This would just pass the energetic test of section 3.5, so that Press' mechanism is energetically possible if a large fraction of the wave energy is spent in mixing of the stratification (rather than being lost by radiative damping, for example). This is uncertain, and since estimates on the amount of wave energy generated are also rather unreliable, it seems that a more detailed analysis is required.

4.2. Mixing due to differential rotation.

In both Schatzman's (1977) and Zahn's (1983) scheme (discussed below) differential rotation is assumed to cause hydrodynamic instability of some form which breaks down into turbulence, leading to mixing. For a given total angular momentum, a differentially rotating star contains more rotational energy than a solidly rotating one. This energy difference can be dissipated by internal processes in the star. For "well behaved" angular momentum distributions inside the star, the energy difference is of the same order as the rotational energy of the uniformly rotating star, or less. For stars with significantly evolved cores, the energy needed for mixing of the core is much larger than their rotational energy content (except perhaps a few very rapidly rotating A stars). Unless differential rotation is regenerated by internal processes in the star (as in Zahn's scheme) the energy argument in 3.5 shows that differential rotation would not be energetic enough to mix the core. In a different way, this is also apparent from the conditions for instability of the various hydrodynamic instabilities discussed in section 5. For slow rotation (in the sense $\Omega \ll N_\mu$), several instabilities are suppressed. In those that survive, the only unstable directions of fluid motion are almost horizontal (as in the case of the ABCD instability), or the instability is effective only over a small range in depth (as for the baroclinic instability near a μ-jump). Since hydrodynamic instability requires that energy can be extracted by exchange of fluid parcels somewhere within the flow, this connection with the energy argument of 3.5 is not surprising.

4.2.1. Schatzman's scheme.

As shown first by Schatzman (1962), the torque exerted by a magnetic stellar wind is quite effective in spinning down a star, thus explaining the low rotation rates of late type stars (see also Weber and Davis, 1967, Mestel, 1968a, Mestel 1968b). The convection zone, being highly viscous due to convective turbulence, rotates at the surface rate (except for certain processes which redistribute angular momentum inside the convection zone and maintain a steady state of differential rotation). In Schatzman's (1977) proposal, the shear between the more rapidly rotating zone and the convection zone is unstable and leads to turbulence which transports the angular momentum from the core to the surface. Since a prediction of the level of turbulence and its transport efficiency is difficult, Schatzman and Maeder (1981) modeled the turbulence by an effective diffusivity which is larger than the miscroscopic viscosity by a constant factor, called R*. A value of Re* between 100 and 200 reproduces the Li, Be and B abundances of the sun (Vauclair et al., 1978). Schatzman and Maeder (1981) found that a quite similar value can explain the enhanced abundance of ^{3}He at the solar surface. This ^{3}He is the product of partial hydrogen burning in the outer parts of the core.

Turbulent mixing associated with spindown, then, seems a promising possiblity for explaining the Li, Be, and ^{3}He effects. Doubts, however, can be raised about the proposed constancy in time of Re*. Observations of the rotation rate of stars vs. age (e.g. Soderblom, 1983) show that the rate of spindown must have been much faster initially than it is now. Turbulence generated by differential rotation would then also be much higher initially, which suggests that Re* should decrease with time. In fact, since the total angular momentum of the star is finite, the total amount of mixing that can be achieved by spinning down the star completely is also finite, and it must be checked whether this amount is sufficient to explain each of the observed effects. We return to this question in section 6.

4.2.2. Zahn's scheme.

The Eddington-Sweet-Vogt circulation transports angular momentum and thereby is a continuous source of differential rotation, unrelated to that caused by spindown. The <u>energy</u> contained in this differential rotation derives, via the circulation, from the luminosity of the star, so that a much larger energy reservoir is in principle available for mixing. Of course, the process can function only if an ESV circulation exists, so that it will not work in the region of strong composition gradients in the core. As argued by Zahn (1983) the circulation would lead to mixing along the following steps. Since the circulation is slow, only a small amount of differential rotation is generated, which would be insufficient to cause shear instabilities that mix <u>across</u> equipotentials (cf. section 5). On a given equipotential however, arbitrarily small differences in rotation rate could become unstable to some form of shear instability. The turbulence created this way would be two-dimensional, the velocity component perpendicular to the equipotential being zero except for a small amount of superposed internal gravity waves. This is due primarily to the very strong influence of the coriolis force on the motions, as in the case of the earth's atmosphere (but in a more extreme way). The turbulence cascades towards smaller scales, until the turnover time of an eddy becomes comparable to the rotation period. From there on, the motion becomes 3-dimensional. Thus it is this small-scale tail of the 2-d turbulence which will cause mixing. Zahn calculates the mixing rate for this process, and finds that the typical turbulent diffusion time scale across the star is of the same order as the ESV circulation time scale. Thus, the process would have some effect in stars which rotate so rapidly that their ESV circulation time scale is shorter than their main-sequence life (some A stars, mostly). For the sun, unfortunately, the effects would be small (cf. section 3.2).

5. HYDRODYNAMIC INSTABILITIES.

The instabilities reviewed here have also been summarized by Zahn (1983); we refer to this text for a more detailed discussion.

An unstratified, viscous fluid subject to a sheared flow can be unstable by a number of mechanisms, generally called "shear instability". Relatively few analytical results on the location of stability boundaries are available, due, among other things, to the highly nonlinear nature of the instabilities. Comprehensive discussions can be found in Drazin and Reid (1981) and Joseph (1976). In a star, the situation is complicated in essential ways by effects of

- rotation
- the stable thermal stratification
- large thermal diffusion coefficient (low Prandtl number)
- stable composition gradients.

As a result, several distinct kinds of instability are possible. Of particular importance is the effect of stratification. Let N be the buoyancy frequency (the frequency of adiabatic oscillation of a fluid parcel due to gravity), given by

$$N^2 = N_\mu{}^2 + N_T{}^2 ,$$

where

$$N_\mu{}^2 = -g\partial_r \ln\mu ,$$

$$N_T{}^2 = g/H\,(\nabla_a - \nabla),$$

and μ is the mean atomic weight per particle, g the acceleration of gravity, H the scaleheight $H=-dr/d\ln P$, $\nabla_a = 1-1/\gamma$ the adiabatic gradient and $\nabla = d\ln T/d\ln P$ the actual temperature gradient. We call N_μ the compositional buoyancy frequency and N_T the thermal buoyancy frequency. The effect of the stratification can be measured by the dimensionless parameter Ω/N. If $\Omega/N \gtrsim 1$, the effect of the stratification is weak, and the various instabilities occur, roughly, under the same conditions and are not so clearly separated. The case $\Omega/N \gtrsim 1$ can occur in a significant part of the star only if it rotates near its breakup rate. We will call this the rapidly rotating case. For <u>slowly</u> rotating stars, Ω/N can be of order unity either very close to its center (because the acceleration of gravity vanishes there) or close to a convection zone. For the situation of interest here, namely instability in a significant part of a slowly rotating star, we may assume

$$\Omega/N \ll 1.$$

The effects of stratification are then strong, and one must distinguish between instability due to <u>radial</u> and <u>horizontal</u> gradients in rotation rate. Since purely horizontal motions do not experience the stabilizing effects of the stratification, it is conceivable (Zahn, 1974, 1975) that even a very weak dependence of rotation rate on latitude may lead to two-dimensional (horizontal) turbulence. If the differential rotation is indeed weak, the turbulence will be dominated by the coriolis force (geostrophic turbulence). Such turbulence was discussed by Zahn (1983).

Instability of radial gradients in the rotation rate, for example due to the torque exerted by the magnetic stellar wind, behaves rather differently. In the limit (1), only sharp gradients in the rotation rate can become unstable to adiabatic motions. Such a sharp gradient can be treated in a plane parallel approximation, for which the Richardson criterion applies. It says that instability is possible only if

$$Ri \equiv N^2/(\partial_r U)^2 < \frac{1}{4} \qquad (2)$$

(Miles, 1961; Howard, 1961), where the shear rate $\partial_r U$ corresponds to $r\partial_r\Omega$. With $N \gg \Omega$ this shows that this form of instability occurs only for strong gradients of Ω, and is therefore limited at best to very localized areas in the star. For motions occuring on sufficiently small scales the thermal diffusion time scale can be less than their turnover time, so that the stabilizing effect of the thermal stratification is smaller. A linear stability analysis in the presence of a thermal stratification and thermal diffusion was done by Jones (1977). An approximate theory for turbulence in this case was formulated by Townsend (1958, see also Turner, 1973), and applied to the astrophysical case by Zahn (1974, 1975, 1983). Including the effect of a composition gradient as well as a thermal stratification, Zahn (1983) finds that instability will occur if the approximate condition

$$r\partial_r \ln\Omega > \max\,[(\tfrac{\nu}{\kappa}\,Re^c)^{1/2}\,N_T/\Omega,\ N_\mu/\Omega] \qquad (2)$$

is satisfied, where ν is the viscosity, k the thermal diffusivity (in cm^2s^{-1}), and Re^c the critical Reynolds number. If the rotation $\Omega(r)$ does not contain a turning point, Re^c is probably of the order 10^3, but in the neighborhood of a turning point it could be much smaller. Since ν/κ, the Prandtl number is very small, instability in a purely thermal stratification (N_μ=0) is possible even for fairly high values of N_T/Ω. In evolved parts of the star, however, N_μ/Ω will typically be very large, so that instability is suppressed according to condition (2).

In addition to the barotropic shear instability governed by Zahn's condition (2), a more subtle form of shear instability is possible if the rotation rate is not constant on cylinders coaxial with the rotation axis, i.e. if the star is <u>baroclinic</u>. This form of instability plays a major role in the earth's atmosphere, and has been analysed in detail in the geophysical context (e.g. Pedlosky, 1979). Baroclinic instability has been invoked as a universal source of turbulence in stars by Tassoul and Tassoul (1982). This has been disputed by Spruit and Knobloch (1983) who showed, on the basis of a more detailed analysis, that the known conditions for baroclinic instability exclude its presence in slow rotators, except for very narrow regions near the core of a star, near a convection zone or near a sharp jump in composition.

In addition to the nonaxisymmetric shear-type instabilities discussed above there exist two axisymmetric instabilities of the double diffusive variety (for reviews of double diffusive instabilities, see Turner, 1973, Huppert and Turner, 1981, Acheson, 1978). These are the Goldreich-Schubert-Fricke (GSF) instability (Goldreich and Schubert, 1967, Fricke, 1968, McIntyre, 1970, Knobloch and Spruit, 1981) and the ABCD instability (McIntyre, 1970, Shibahashi, 1980, Knobloch and Spruit, 1982). The GSF instability occurs if

$$r\partial_r \ln\Omega > 2(\nu/\kappa)^{1/2}\,N_T/\Omega\,(1 + \kappa/\kappa_4\,N_\mu{}^2/N_T{}^2)^{1/2}\ , \qquad (3)$$

and the ABCD if

$$r\partial_r \ln\Omega > 2\sqrt{2}\,(\nu/\kappa)^{1/2}\, N_T/\Omega\, f(N_\mu) \qquad (4)$$

(Knobloch and Spruit, 1982), where f is a weak function of N_μ, and κ_4 the diffusivity of ^{4}He (more generally the dominant element determining the compositional buoyancy frequency N_μ). In unevolved parts of the star, both GSF and ABCD instability can occur for low rotation rates and moderate gradients in rotation rate. Since the factor κ/κ_4 in Eq.(3) is large (of the order 10^6 in the sun) even a weak composition gradient has a strong stabilizing effect on the GSF instability (Goldreich and Schubert, 1968). The ABCD instability is much less sensitive to a μ-gradient. This is related to the fact that the fluid motions in this instability are nearly horizontal, so that they avoid doing work against the stable μ-gradient. This suggests that the instability is probably not very effective in mixing the μ-gradient, and this would be consistent with the implications of the energy argument of section 3.5. An estimate of the nonlinear development of this instability would be necessary to determine precisely how strong its mixing effect can be.

We can summarize this discussion of hydrodynamic instabilities as follows. In a rapidly rotating star (near critical rotation) each of the instabilities mentioned can operate. The most effective ones are probably barotropic and baroclinic shear instabilities. In slowly rotating stars, we must distinguish between parts of the star in which the μ-gradient is very weak (in the sense $N_\mu{}^2/N_T{}^2\ \kappa/\nu \lesssim 1$). A number of instabilities can operate here, including shear instability according to Zahn's condition (2), the GSF and the ABCD instability. These instabilities may therefore play an important role in the Lithium depletion, as proposed by Schatzman (1977). In the evolved part of the core ($r \lesssim 0.5R_\odot$) only two instabilities are known: shear instability on horizontal surfaces driven by a latitude depedence of the rotation rate (Zahn, 1975, Watson, 1980) and the ABCD instability. The amount of mixing by these mechanisms is uncertain but on energetic grounds the prospects are not overly promising.

6. EVOLUTION CALCULATIONS WITH MIXING.

Evolution of the sun with partial mixing and angular momentum transport by hydrodynamic instabilities has been studied by Endal and Sofia (1981), Schatzman and Maeder (1981), Law et al. (1984, 1985), D'Antona and Mazzitelli (1984). Endal and Sofia (1981) included the torque exerted by the solar wind as the source of differential rotation, and meridional circulation, shear and GSF instabilities as transport mechanisms. Since these mechanisms are easily suppressed by μ-gradients, their model develops a rapidly spinning nearly unmixed inner core surrounded by a slowly rotating mixed outer core. The rotation rate of the core in this model is much too high to be compatible with the measurements of Duvall and Harvey (1984), or even with the upper limits on oblateness and quadrupole moment of the sun. Calculations with an enhanced diffusivity $\nu_t = Re^* \nu$, where ν is the microscopic viscosity were done by Schatzman and Maeder (1981). Assuming that Re* is constant in time, values of Re*=50-200 yield satisfactory values for the neutrino flux and the ^{3}He and Li surface abundances. By using eq.(2) to model the dependence of Re* on radius Schatzman (1985) obtained an even closer fit. The angular momentum transported by the enhanced viscosity in these models is not enough to explain the data on the internal rotation of the sun, the core rotates 10-30 times too fast (Schatzman, private communication; Law et al. 1984, 1985). Law et al. (1984, 1985) point out that the mixing from differential rotation-induced instabilities should be much stronger initially when the spindown rate is large, than in the present sun, i.e. Re* is expected to be a rapidly decreasing function of time. The Lithium depletion depends on the total amount of mixing that takes place over the life of the star, but the neutrino flux is affected only by mixing which takes place later when an evolved core has developed. The effect on Lithium depletion relative to that on the neutrino flux is therefore much larger for such a time dependent mixing than for constant mixing. Also, the total amount of mixing available is limited by the total initial angular momentum of the star, and one may wonder whether this amount is enough to reduce the neutrino flux sufficiently. Law et al. calculated models under the assumption that the core is mixed by the ABCD instability since this is the most readily occuring instability driven by radial differential rotation. They consider two recipes for mixing by the ABCD instability, both of which may be somewhat optimistic in view of the energy argument presented in 2.5. The most efficient mixing is obtained by assuming that the instability transports angular momentum so quickly that the instability brings the rotation rate $\Omega(r)$ close to marginal stability. The rotation curve $\Omega(r)$ for the present sun under this assumption was given by Spruit et al. (1982). The rotation rates in this model are clearly too high compared with the observed rotation rates of Duvall and Harvey. The reduction of the neutrino flux for this model is significant but not quite large enough while the Lithium depletion is far too large, due to the effect discussed above. The conclusion from the calculations of Law et al. is that no hydrodynamic instability is known which can simultaneously explain the Lithium and neutrino data, or which explains the low rotation of the interior.

7. DISCUSSION, PLAUSIBLE ALTERNATIVES.

On the basis of the foregoing, it seems unlikely that a single transport mechanism can account for the neutrino flux, the Lithium depletion and the internal rotation of the sun, in spite of the initial success of Schatzman's suggestion. For mixing in the inner core, the two mechanisms which can not yet be ruled out at present are the ^{3}He instablity and Press's gravity-wave mixing but both require more work. Recalling that the present p-mode oscillation data (Christensen-Dalsgaard, 1984) as well as the preliminary identifications of g-mode oscillations by the Stanford group (Delache and Scherrer, 1984) favor an almost unmixed model (see however Berthomieu et al. 1984), one may wonder whether it is worth the effort to pursue these possibilities much further. For possibilities to explain the neutrino flux without mixing, see the review by Bahcall et al. (1981). The Li depletion and ^{3}He enhancement at the solar surface indicate real mixing in the outer core, and a number of possible mechanisms have been identified, including shear instabilities. More work is required to determine the amount of mixing to be expected from these mechanisms.

A key observation is the low rotation rate of the interior, since it is hard to explain with any of the proposed mixing mechanisms, especially in the evolved part of the core. On the other hand, as discussed in 3.3 only a weak magnetic field is sufficient to exert the internal torques needed to keep the sun rotating almost uniformly. For example the field created by the winding up of a 10μG primordial field by a moderate degree of differential rotation would be sufficient (Mestel, 1961).

REFERENCES

Acheson, D.J., 1978, Phil. Trans. Roy. Soc. London **289A**, 459.

Bahcall, J.N., Huebner, W.F., Lubow, S.H., Parker, P.D., Ulrich, R.K., 1982, Rev. Mod. Phys. **54**, 767.

Berthomieu, G., Provost, J., Schatzman, E., 1984, Nature **308**, 254.

Bienaymé, O., Maeder, A., Schatzman, E., 1983, Astron. Astrophys. **131**, 316.

Biermann, L., 1950, Z. Naturforschung., **5a**, 65.

Boesgaard, A.M., and Tripicco, M.J., 1986: Ap. J. **303**, 724.

Brown, T.M., 1985: Nature **317**, 591.

Busse, F.H., 1981, Geophys. Astrophys. Fluid Dyn., **17**, 215.

Cayrel, R., 1984, in A. Maeder and A. Renzini, eds., Observational tests of the stellar evolution theory (IAU symp. 105), D. Reidel, Dordrecht, p. 533.

Cayrel, R., Cayrel de Strobel, G., Campbell, B., Däppen, W., 1984, Astrophys. J. **283**, 205.
Christensen-Dalsgaard, J., Dilke, F.W.W., Gough, D.O., 1974, Mon. Not. R. Astron. Soc. **169**, 429.
Christensen-Dalsgaard, J., 1984, in T. Guyenne, ed., ESA Special publication SP 220, European Space Agency, Noordwijk, p.3.
D'Antona, F., Mazzitelli, I., 1984, Astron. Astrophys. **138**, 431.
Delache, P., Scherrer, P.H., 1983, Nature **306**, 651.
Dilke, J., Gough, D.O., 1972, Nature **240**, 262.
Drazin, P., Reid, W.H., 1981, Hydrodynamic Stability, Cambridge University Press, Cambridge.
Duncan, D.K., 1981, Astrophys. J. **248**, 651.
Duncan, D.K., Jones, B.F., 1983, Astrophys. J. **271**, 663.
Duvall, T.L., Harvey, J.W., 1984, Nature **310**, 19.
Duvall, T.L., Dziembowskie, W.A., Goode, P.R., Gough, D.O., Harvey, J.W., Leibacher, J.W., 1984, Nature **310**, 22.
Endal, A.S., Sofia, S., 1981, Astrophys. J. **243**, 625.
Fricke, K.J., 1968, Z. Astrophys. **68**, 317.
Genova, F., Schatzman, E., 1979, Astron. Astrophys. **78**, 323.
Goldreich, P., Schubert, G., 1967, Astrophys. J. **150**, 571.
Howard, L., 1961, J. Fluid Mech. **10**, 508.
Huppert, H.E., Spiegel, E.A., 1977, Astrophys. J. **213**, 157.
Huppert, H.E., Turner, J.S., 1981, J. Fluid Mech. **106**, 299.
Joseph, D.D., 1976, Stability of Fluid Motions, vols. I and II, Springer, New York.
Knobloch, E., Spruit, H.C., 1982, Astron. Astrophys. **113**, 261.
Knobloch, E., Spruit, H.C., 1983, Astron. Astrophys. **125**, 59.
Law, W.-Y., Knobloch, E., Spruit, H.C., 1984, in A. Maeder and A. Renzini, eds., Observational tests of the stellar evolution theory, D. Reidel, Dordrecht, p. 523.
Law, W.-Y., Spruit, H.C., Knobloch, E., 1985, in preparation.
McIntyre, M.E., 1970, Geophys. Fluid Dyn. **1**, 19.
Mestel, L., 1953, Mon. Not. R. Astron. Soc. **113**, 716.
Mestel, L., 1961, Mon. Not. R. Astron. Soc. **122**, 473.
Mestel, L., 1968a, Mon. Not. Roy. Astron. Soc. **138**, 359.
Mestel, L., 1968b, Mon. Not. Roy. Astron. Soc. **140**, 177.
Mestel, L., 1984a, Astron. Nachr. **305**, 301.
Mestel, L., 1984b, in A. Maeder and A. Renzini, eds., Observational tests of the stellar evolution theory, D. Reidel, Dordrecht, p.513.
Mestel, L., Moss, D.L., 1983, Mon. Not. Roy. Astron. Soc. **204**, 557.
Michaud, G., 1986, Astrophys. J. **302**, 650.
Miles, J.W., 1961, J. Fluid Mech. **10**, 508.
Osaki, Y., 1982, Publ. Astron. Soc. Japan **34**, 257.
Pedlosky, J., 1979, Geophysical Fluid Dynamics, Springer, New York.
Press, W.H., 1981, Astrophys. J. **245**, 286.
Rosner, R., and Weiss, N.O., 1985, Nature **317**, 790.
Roxburgh, I.W., 1984, in A. Maeder and A. Renzini, eds., Observational tests of the stellar evolution theory, D. Reidel, Dordrecht, p.517.
Schatzman, E., 1962, Ann. d'Astrophys. **25**, 18.
Schatzman, E., 1977, Astron. Astrophys. **56**, 211.
Schatzman, E., 1985, in preparation.
Schatzman, E., Maeder, A., 1981, Astron. Astrophys. **96**, 1.
Shibahashi, H., 1980, Publ. Astron. Soc. Japan **32**, 341.
Skumanich, A., 1972, Astrophys. J. **171**, 565.
Soderblom, D.R., 1983, Ap. J. Supp. **53**, 1.
Spruit, H.C., Knobloch, E., 1983, Astron. Astrophys. **132**, 89.
Spruit, H.C., Knobloch, E., Roxburgh, I.W., 1983, Nature **304**, 320.
Tassoul, J.-L., Tassoul, M., 1982, Astrophys. J. Supp. **49**, 317.
Tassoul, J.-L., Tassoul, M., 1984, Astrophys. J. **279**, 384.
Townsend, A.A., 1958, J. Fluid Mech. **4**, 361.

Turner, J.S., 1973, Buoyancy effects in fluids, Cambridge University Press.
Vauclair, S., Vauclair, G., Schatzman, E., Michaud, G., 1978, Astrophys. J. **223**, 567.
Vauclair, S., Vauclair, G., 1982, Ann. Rev. Astron. Astrophys. **20**, 37.
Watson, M., 1980, Geophys. Astrophys. Fluid Dyn. **16**, 285.
Weber, E.J., Davis, L. Jr., 1967, Astrophys. J. **148**, 217.
Zahn, J.-P., 1974, in Stellar Stability and Evolution, eds. P. Ledoux, A. Noels and A.W. Rogers, Reidel, Dordrecht.
Zahn, J.P., 1975, Mem. Soc. Roy. Sci. Liege 6^e Serie, **8**, 31.
Zahn, J.-P., 1983, in A.N. Cox, S. Vauclair and J.-P. Zahn, eds., Astrophysical Processes in Upper Main Sequence Stars, Geneva Observatory, p. 253.

Part III

Single and Binary Stars

Models of Star Formation

W.M. Tscharnuter

Institut für Astronomie, Universität Wien,
Türkenschanzstr. 17, A-1180 Wien, Austria

The collapse of spherically symmetric protostars has been reinvestigated on the basis of new numerical tools and techniques that have been recently developed. In addition, a much more consistent set of thermodynamic functions (particularly for molecular hydrogen) and a substantially improved version of the equation of state have been used. Protostellar cores are found to be always on the verge of dynamical instability due to the fact that the gas is cool and dense there, and the adiabatic exponent $\Gamma_1 = (\partial \ln P / \partial \ln \rho)_{ad}$ varies around the critical number 4/3. During the overall collapse of a Jeans-unstable cloud fragment, which lasts 3-4 initial free-fall times, many cores form and undergo subsequent explosive reexpansion in a quasi-periodic way, the cycles becoming shorter and shorter. The final contraction to a (pre-) main-sequence star takes place on time scales of a few thousand years. It starts out with much higher mean densities, yielding an optical depth for the fragment of the order of unity and, as a consequence, higher temperatures in the core which now stabilizes. The main accretion phase appears to be nothing but a last-minute-effect, the probability to observe 'classical' protostars is thus very low.

INTRODUCTION

The development of a detailed theory of stellar formation is one of the most intriguing astrophysical problems. It is by far not complete at present. As the mathematical difficulties are rather big and the physics involved is very complex, only the simplest case of spherical symmetry has been investigated to a certain degree of completeness. Some progress has been made with axially symmetric (2D-) models, general 3D-flows exhibiting density contrasts greater than 9-10 orders of magnitude evade a proper numerical description for reasons of accuracy; yet another ten orders of magnitude more are encountered within protostellar collapse flows. The present state of art in this field of research has very recently been discussed by Boss [1]. So I should like to restrict myself to the presentation of new results mainly pertaining to the spherically symmetric (1D-) collapse which have not yet been discussed in the literature and are likely to demand a revision of a substantial part of the 'classical' theory. 2D-calculations which are currently being carried out confirm the earlier finding concerning the instability of the core in a rotating turbulent protostellar cloud (cf. [2]). The cyclic formation and disruption of the core seems to be a general physical feature of collapsing gas-dust clouds if angular momentum has been removed from the central parts in some way or another. The long-lasting main accretion phase suggested by former calculations may well be an oversimplification of the true situation.

Stars are formed from the rather dispersed interstellar medium, young stellar objects are always closely associated with interstellar clouds of gas and dust. Theory has to explain why and how the transition from the cloud state to the highly condensed pre-main-sequence stars or, in case of more massive objects, even directly to zero-age-main-sequence stars takes place. In particular, the intermediate stages are not easily accessible for observations, but theoretical model sequences could provide us with the information which phenomena (luminosity, spectral appearance, variability ...) ought to be expected and what we should really look for. Among the numerous infrared point sources recently discussed (cf. [3],[4]) there might be several genuine protostellar objects, i.e. star-like objects which are about to rapidly accrete a substantial fraction of their mass; however, up to now every infrared source referred to as a 'protostar' in the first place later turned out to be a heavily obscured star with mass outflow rather than mass infall.

The equations which describe protostellar evolution are intrinsically nonlinear, they express the complex interaction between gasdynamics, thermodynamics (ionization, dissociation ...), radiation, and self-gravitation, only to mention the very indispensable processes! Likewise of great importance are magnetic fields, the chemistry in the gas phase and on the surfaces of small dust particles as well as the growth of the dust grains by agglomeration, thermonuclear reactions. In order to combine all these effects in a consistent way the preparation of more and more refined numerical models is of utmost importance. Large-scale subsonic and supersonic motions of the gas are encountered, shock fronts are common phenomena in self-gravitating flows. Extremely different time and length scales varying in between

6-9 orders of magnitude pose stringent conditions on any numerical method to be successfully applied. There is still no really comfortable code available that would allow us to do more extensive parameter studies, although great progress has been made by introducing adaptive grids into implicite numerical schemes [5]. Thus, what have we learnt at all about star formation from numerical experiments?

THE CLASSICAL PICTURE

The major breakthrough in the theory of star formation came along with the work of Larson [6] in 1969. He was the first to fully recognize the importance of the extreme non-homology of collapse flows and the occurence of the dynamical instability due to thermal destruction of molecular hydrogen in the first quasihydrostatic core. Starting out with a cloud fragment on the verge of being Jeans-unstable only a few percent of a solar mass (independent of the total mass involved) are contained within a 'stellar' core after the collapse has stopped, the rest of the material is gradually collected by the core during 3-4 initial free-fall times. In this long-lasting accretion phase the stellar 'embryo', separated by a strong accretion shock from its almost freely falling envelope, grows up into a (pre-) main-sequence star as the envelope itself gets exhausted from matter.

Honest time-dependent hydrodynamical model sequences covering the whole evolution from the Jeans-unstable fragment to the final star are extremely difficult to obtain without simplifying assumptions, even in the spherically symmetric (1D-) case. On that account only a few calculations of Larson's type have been published ([7], [8], [9], [10]). Because of the technical difficulties inherent to the problem any conventional hydrodynamical textbook code would break down shortly after the second collapse and never find back to larger time steps (sometimes more than a billion Courant steps) needed in order to cover the main accretion phase. Larson himself used a rather crude numerical method and was forced to <u>assume</u> mechanical equilibrium to hold for protostellar cores and the accretion shock to be isothermal. Remarkably enough, Larson's results have been well confirmed by any other calculation for which those stringent assumptions could be abandoned. However, the crucial question relates to the dynamical behavior of the stellar core: is there always enough <u>physical</u> damping of the oscillations which the core undergoes after the bounce? As a matter of fact, great efforts have been made to 'stabilize' the core, mostly by an increase of <u>numerical</u> damping (coarse spatial grid, large time steps, completely backward time differences), since there has never been much interest in the details of the internal structure of the stellar embryo at its age of a few days or weeks. Nevertheless, a more detailed investigation of the cool and dense protostellar core interiors could provide us with new insights into the global phenomenon of star formation. There is indeed a hint from 2D-calculations [2] that the tiny protostellar cores are quite 'fragile' structures which can easily undergo not only oscillations but also explosive expansion.

DYNAMICAL INSTABILITY

After formation of the (first) optically thick core the former isothermal evolution quickly turns over to an adiabatic one. At a central density of about 10^{-12} g/cm^3 the accretion time scale becomes shorter than the Kelvin-Helmholtz time of the core. There is only a thin layer downstream from the accretion shock where energy transport by radiation and/or convection dominates. At densities greater than 10^{-8} g/cm^3 the temperatures rise above 2000 Kelvin. Molecular hydrogen then dissociates and the adiabatic exponent $\Gamma_1 = (\partial \ln P / \partial \ln \rho)_{ad}$ becomes smaller than 4/3. This is the critical value below which the equilibrium of the (spherically symmetric) core is dynamically unstable (cf. [11]).

The instability sets in while the (first) core is contracting due to mass accretion and, hence, triggers a rapid collapse-like dynamical <u>contraction</u>. Within less than one year the central density reaches values above 0.01 g/cm^3 . The second core forms when Γ_1 becomes again greater than 4/3. However, the temperatures there remain far below 10 Kelvin so that (pressure-) ionization, Coulomb effects and even partially degenerate electron gas contribute to the pressure and internal energy of the material. The resulting adiabatic exponent is found to be close to the critical 4/3 throughout the core. As a consequence, the bounce after the second collapse gives rise to oscillations of large amplitudes, unless one manages to apply a sufficiently effective damping mechanism, as mentioned above. This is exactly what has been done in all existing model calculations in order to save computer time.

But woe, if after the bounce the thermodynamical properties of the adiabatically cooling gas in the expanding core were such that $\Gamma_1 < 4/3$ again! Both recombination processes and the re-formation of molecular hydrogen, the inverse processes of ionization and dissociation, would then trigger an <u>explosive expansion</u> of the stellar core, leading to its complete destruction. If this mechanism works, the quasi-stationary main accretion phase, as postulated by the 'canonical' theory, dissolves into a series of non-stationary oscillatory cycles. Relatively long-lasting, highly dilute states of protostellar matter will then alternate with short-lived, episodic appearances of condensed stellar cores.

ON THE EXISTENCE OF 'OSCILLATORY CYCLES'

Numerical model sequences of spherically symmetric protostellar collapse have been reinvestigated by collaborators of the Vienna Astrophysical Seminar (VAS) during the past four years. A new radiation-hydrocode (implicit, first order donor cell) in conservation form (cf. [12]) with an adaptive grid according to [5] has been written. The equation of state and the thermodynamic functions of a Pop. I mixture, with the density and internal energy as the primary variables, have been calculated by G. Wuchterl (VAS). Most useful reference data for pure hydrogen have been kindly placed to our disposal by W. Huebner (Los Alamos). The interpolation routines (two-dimensional rational splines with smoothing parameter) for the gas pressure and the temperature as functions of the gas density and internal energy as well as for the opacity have been written by M. Stoll (VAS). For modeling the radiation field the time dependent Eddington (diffusion) approximation and, for comparison purposes, the radiative conduction equation have been used. Various test calculations for quite different opacity values did not change the qualitative picture; the anisotropy (Eddington) factor thus does not seem to play a decisive role for the physics of the core and, for the sake of saving computer time, has been set equal to 1/3 throughout. A respective modification of the code has almost been finished (M. Balluch, VAS). A more severe shortcoming has been found in neglecting convective heat transport, in particular during the main accretion phase. However, it is not clear whether convection will also be important as an effective damping mechanism of core oscillations in the early phases. An incorporation of a suitable time dependent formulation into the code is currently being on the way (M.G. Fernandes, VAS).

The conjectures advanced in the previous section have been fully confirmed by a series of test calculations. We have found that the variation of the adiabatic exponent leads to the onset of the dynamical instability, provided the heat transport away from the accretion shock region is not too high. In case of the Eddington approximation the accretion shock will move with a higher velocity relative to the infalling gas, and more entropy will be deposited in the shocked gas than in the case of radiative conduction, where the isothermality of the accretion shock is forced. This flow feature comes along with the general expansion of the core after the bounce. According to the numerical results the Eddington approximation makes the expansion continue until the destructive dynamical instability sets in, whereas radiative conduction is able to stop the core expansion just in time. In the 2D-case (cf. [2]) the expansion of the (turbulent) core is driven by redistribution of angular momentum, the entropy needed is then generated by turbulent friction. So the instability occurs even if the radiative conduction assumption is adopted. Very recent 2D-calculations show that turbulent heat conduction is also not sufficient to stabilize the core. A more detailed discussion of these computations currently being carried out will be given elsewhere.

The existence of oscillatory cycles has been established for a wide range of initial conditions. It turns out that the occurrence of the instability is completely independent of the initial values of the density (homogeneously distributed or centrally precondensed clouds) as long as the cloud is optically thin to infrared radiation. There is also no influence of the total mass contained in the fragment, at least in the range between 1 and 100 solar masses. Fig. 1 displays the typical variation of the central density with time for a $3M_{\odot}$ cloud which has already undergone more than a dozen cycles since the first

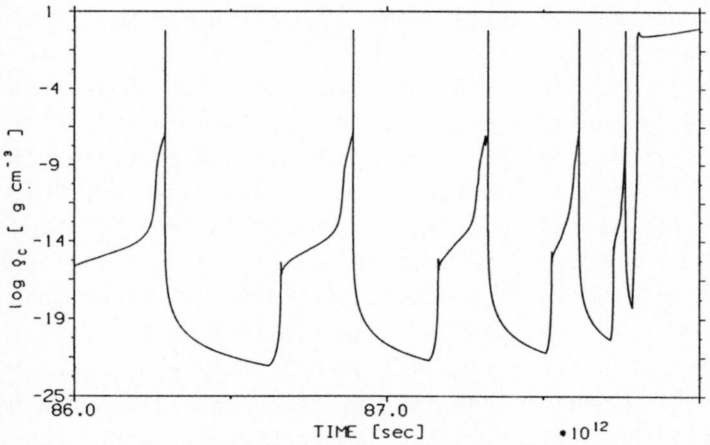

Fig. 1. Central density as a function of time for the last 'oscillatory cycles' before the final accretion

free-fall time (2.1 10^{13} sec). The speeding up of the cycles by a factor of roughly 150 reflects the total increase of the mean density by a factor of about 20000. The final accretion phase lasts only about 10^4 yr, which is just a few tenths of a percent of the total evolution time scale.

The calculation was stopped when about 2/3 of the total mass was in the core. At that point convection would have become already rather important. In addition, the central density was about to rise above 100 g/cm^3, which is the upper limit chosen for our tables. But the most important result is the fact that the stability of the core will be achieved, only if the initial density is several orders of magnitude higher than the critical Jeans-density with respect to a given initial temperature (typically 8-14 Kelvin). At the beginning of this very 'last' collapse directly leading to a star, the optical depth ought to be of the order of unity or even higher. As a consequence, the non-homology of the collapse weakens, the temperature of the (more massive) core becomes higher, and the adiabatic exponent remains well above 4/3. These high mean density for the initial condition should be compared with earlier assumptions made by Hayashi and collaborators [13].

CONCLUSIONS

In the new picture of star formation as discussed above the final accumulation of the star is a last-minute-event at the very end of the global collapse of a gravitationally unstable cloud fragment, when the mean density has already become quite high. The important practical implication of this theoretical finding is the low probability for detecting a 'classical' protostar in its main accretion phase; the probability to observe a protostar with an accreting core during its oscillatory phase is almost zero! Most of the time a core doesn't exist at all, the typical average velocities, both infall and outflow, are of a few kilometers per second on linear scales of several 10^{16} cm. The bulk of the galactic infrared point sources ought to be nothing but (pre-) main-sequence stars which are heavily obscured by matter that could not be accumulated, e.g. because of angular momentum effects and/or the onset of a strong stellar wind and/or radiation pressure for more massive objects. These predictions fit well the empirical result that mass outflow of newly born stars is the common feature rather than mass infall.

ACKNOWLEDGMENTS

I thank Prof. Kippenhahn for his continuous interest in my work on star formation over the past fifteen years. Without his stimulating support and encouragement I would not have been able to make any major progress in this field of research.

Most of the calculations reported on in this paper have been carried out on the CRAY-1(XMP) of the Institut für Plasmaphysik, Garching. Extended testruns were made on the VAX 11/750 of the Institut für Astronomie in Vienna. The work was partly supported by the Austrian 'Fonds zur Förderung der Wissenschaften', project number P5551.

REFERENCES

[1] Boss, A.P. 1986, Theory of Collapse and Protostar Formation, in 'Proceedings from the Summer School on Interstellar Processes', ed. D. Hollenbach and H. Thronson, D. Reidel, preprint.

[2] Morfill, G.E., Tscharnuter, W.M., and Völk, H.J. 1985, Dynamical and Chemical Evolution of the Protoplanetary Nebula, in 'Protostars and Planets II', ed. D.C. Black and M.S. Matthews, The University of Arizona Press, Tucson, p. 493.

[3] Baud, B. et al. 1984, Ap.J. (Letters), 278, L53.

[4] Beichmann, C.A., et al. 1984, Ap.J. (Letters), 278, L45.

[5] Dorfi, E.A., and Drury, L.O'C. 1985, Simple adaptive grids for 1D initial value problems, MPI H - 1985 - v 21, MPI f. Kernphysik, Heidelberg (preprint, to appear in 'Computational Physics', 1986).

[6] Larson, R.B. 1969, Monthly Notices Roy. Astron. Soc., 145, 271.

[7] Appenzeller, I., and Tscharnuter, W.M. 1974, Astron. and Astrophys., 30, 423

[8] Tscharnuter, W.M., and Winkler, K.-H. 1979, Computer Phys. Comm., 18, 171

[9] Winkler, K.-H., and Newman, M.J. 1980a, Ap.J., 236, 201.

[10] Winkler, K.-H., and Newman, M.J. 1980b, Ap.J., 238, 311.

[11] Cox, J.P. 1980, 'Theory of Stellar Pulsation', ed. J.P. Ostriker, Princeton Series in Astrophysics, Princeton, New Jersey.

[12] Winkler, K.-H., and Norman, M.L. 1986, WH80s: Numerical Radiation Hydrodynamics, in 'Astrophysical Radiation Hydrodynamics',

(Proceedings of the NATO ARW held in Garching, FRG, 1982), ed. K.-H. Winkler and M.L. Norman, p.69.

[13] Narita, S., Nakano, T., and Hayashi, C. 1970, Progr. Theor. Phys., 43, 942.

Time Dependent Convection in Stars – a Review of the Theories

N.H. Baker

Max-Planck-Institut für Physik und Astrophysik, Institut für Astrophysik, Karl-Schwarzschild-Str. 1, D-8046 Garching, Fed. Rep. of Germany and Department of Astronomy, Columbia University, New York, NY 10027, USA

Theories that have been proposed to describe thermal convection in a moving atmosphere are described and compared. Most of them are time dependent variations of mixing-length theory and most are found to be quite closely related. The differences and similarities are noted. A brief account of the applications to pulsating-star problems is given.

I. INTRODUCTION

Thermal convection has a role in determining the structure of many stars. It normally arises when the usual radiative mechanism is unable to transport the entire luminosity of a star, and can be responsible for mixing atomic species in regions that are initially inhomogeneous as well as for transporting momentum and energy. The calculation of these effects presents formidable theoretical difficulties, and as a consequence there is uncertainty about the structure of many types of stars in which convection zones are extensive. This is especially true of the cooler stars like the sun, which have deep convection zones in their comparatively opaque outer layers. For stars that are variable in time, the theoretical problems are even more difficult, because the large-scale mean motion of the stellar gas and the smaller-scale turbulent convective motions interact with each other. The attempt to describe the interaction is called "time-dependent convection theory".

Among the first problems of this nature to arise is that of the pulsating variables, the Cepheids and cepheid-like stars. The pulsations of these stars are spherically symmetric, and the physical mechanism that gives rise to their pulsational instability is well understood. The radiation flowing from a star's interior interacts with and is modulated by matter in the relatively opaque outer layers, and when the temperature and density dependence of the radiative absorption coefficients and specific heats in these layers is favorable, this "radiative excitation" can produce an oscillatory instability. But if convection were to carry most of the energy flux in these critical layers, the radiative mechanism would not work. Then the stability would depend on how the convective flux interacts with the pulsation.

It has long been conjectured that convection is important in the cooler Cepheids. Such stars populate a strip in the HR diagram which, at a given luminosity, has a narrow range of color. Many years ago Baker and Kippenhahn (1965) found that although the usual radiative theory predicts quite well the blue edge of the Cepheid strip, it fails completely to define the red edge. But the simplest possible convection theory, the crude mixing-length theory, when applied to static models of Cepheids, indicates that the cooler ones have extensive surface convection zones in just the part of the star containing the seat of the instability mechanism. If one assumes that the fraction of the energy flux which is carried by convection in the static model simply does not interact at all with the pulsation ("frozen-in" convection), one finds (Baker and Kippenhahn 1965) that the cooler Cepheids become less unstable as the effective temperature is decreased. The reason for this is just that the radiative destabilizing mechanism shuts down as less and less of the flux is carried by radiation in the critical part of the star.

This is an incomplete picture. The coolest models become only neutrally stable, and the red edge thus predicted is definitely too red. Even for stars in the center of the strip, mixing-length theory produces models in which the phase relationship between velocity and luminosity variations disagrees with observations (Castor 1971). More recent attempts to model the convection-pulsation interaction have produced better results for the "red-edge problem" (Deupree 1977a,b,c; Stellingwerf 1984a,b,c; Baker and Gough 1979; Gonczi 1982). Some of this work will be described below.

There are many other objects to which a theoretical treatment of time-dependent convection might be applied. The long-period variables are very cool, luminous stars having very deep convective zones. Convection may well be responsible for the pulsational instability of these stars. Some of the early-type nonradial pulsators, like the δ Scuti variables, may have important convection zones. Another interesting class comprises pulsating white dwarfs (ZZ Ceti variables) in which it has been found (Cox et al. 1986) that the assumption of frozen-in convection gives rise to an apparently spurious instability. A better treatment is clearly needed. Even in the sun, which exhibits many modes of very low-amplitude nonradial pulsation, the convection-pulsation interaction is important. At the least, convection slightly alters the pulsational frequencies, which can be very precisely measured (Christensen-Dalsgaard 1982). A time-dependent convection theory is also needed for theoretical studies of nonperiodically varying objects. Examples of these are novae, supernovae, and collapsing protostars.

The purpose of this article is to review briefly most of the theoretical treatments that have been used or proposed for time dependent convection in stars . The derivations given (or indicated) here are not always the same as those of the original authors, and some of the subtleties are omitted, but is is hoped that no great violence is done to the work in this attempt to show how the various methods are related to one another. Although the emphasis is on theoretical developments, some account of the applications made to date is also presented.

II. THE EQUATIONS OF MOTION

Nearly every treatment of time-dependent convection published to date is based on the Boussinesq approximation to the equations of motion, which means that the gas is taken to be incompressible, except for buoyancy effects which are of course essential. In some treatments the assumptions of the Boussinesq approximation are introduced at different points in the development, which is confusing, and it is only by careful examination of the final equations of motion that one can see what has actually been neglected.

The Boussinesq approximation is based on a careful scaling argument and an expansion in small parameters. Strictly speaking, it is probably inapplicable in most stellar situations, but it is internally consistent and is widely used. It underlies mixing-length theory, and most of the treatments to be discussed here can be considered versions of mixing-length theory. I believe that it is more consistent and certainly less confusing to write down the Boussinesq equations at the start, and then develop the models from there.

Helpful discussions of the Boussinesq approximation can be found in Spiegel and Veronis (1960) and Gough (1969), the latter of which will now be summarized briefly. The idea is, first, to eliminate small-scale acoustic modes while retaining other effects of compressibility, and this goal is accomplished by expressing all thermodynamic quantities as the sum of a mean (normally a horizontal mean) and a fluctuating part, and then linearizing the equations in the fluctuating parts. (Nonlinear terms in the velocities are retained.) It turns out that this implies that all fluid velocities are small compared with the sound speed. The continuity equation becomes $\mathrm{div}(\rho\mathbf{u})=0$. Gough (1969) shows that this procedure, called the anelastic approximation, results in a set of equations that are energetically consistent. To obtain the Boussinesq approximation the anelastic equations are themselves expanded in another small parameter, the ratio of the thickness of the convecting layer to a characteristic scale height. This is a more drastic assumption and the resulting equations are strictly valid only in convection zones that are much thinner than those normally encountered in stars. It is implicit in the Boussinesq approximation that spatial derivatives of mean quantities are small compared with those of the fluctuations, and the pressure fluctuation is small compared with those of density and temperature. The continuity equation is that of an incompressible gas.

Despite these rather restrictive assumptions, the Boussinesq equations retain many of the complexities of the full equations of motion. They are also compatible with large-scale acoustic oscillations such as those described by the pulsation equations.

Most investigations have been restricted to the case in which the mean motion is in the radial direction only, as would be the case in spherically symmetric modes of pulsation. In this case it is convenient to write down Lagrangian equations for the mean quantities, retaining an Eulerian description for the convection (fluctuating) quantities. Gough (1969) derives these equations for a plane-parallel atmosphere. In a later paper (1977a) he writes the mean equations in spherical polar coordinates. Rectangular coordinates are retained for the Eulerian equations describing the fluctuations, consistent with the Boussinesq approximation. These equations will be used to illustrate the various approaches that have been taken to modeling Boussinesq convection.

In Gough's (1977a) paper variables are expressed as the sum of mean and fluctuating parts, viz., $T(\mathbf{r},t) = \overline{T(r,t)} + T'(\mathbf{r},t)$.
Mean quantities are averages over a spherical surface and of course depend only on the radial coordinate. The velocity is written as $v_i(\mathbf{r},t) = V(r,t)\delta_{i3} + u_i(\mathbf{r},t)$, the mean velocity having only a radial (x_3) component. The notation $w = u_3$ is used for the radial component of the fluctuating velocity. The complete set of equations is then

$$\frac{\partial}{\partial r}(\bar{p} + p_t) + (3 - \Phi)\frac{p_t}{r} = -g\bar{\rho} \tag{1}$$

where

$$p_t = \bar{\rho}\,\overline{w^2}, \qquad \Phi = \overline{u_i u_i}/\overline{w^2}, \tag{2}$$

$$g = \frac{Gm}{r^2} + \frac{\partial^2 r}{\partial t^2}; \tag{3}$$

$$\frac{\partial m}{\partial r} = 4\pi r^2 \bar{\rho}, \tag{4}$$

$$\bar{c}_p \frac{\partial \bar{T}}{\partial t} - \frac{\bar{\delta}}{\bar{\rho}}\frac{\partial \bar{p}}{\partial t} = -\frac{1}{\bar{\rho} r^2}\frac{\partial}{\partial r}[r^2(\bar{F}_3 + F_c)], \tag{5}$$

where $\delta = -(\partial \ln\rho/\partial \ln T)_p$ and F_i is the radiative heat flux. The Boussinesq expression for the mean convective heat flux is

$$F_c = \bar{\rho}\bar{C}_p\overline{wT'}. \tag{6}$$

It has been assumed here that the turbulence is locally axisymmetric about the radial direction, and the quantity Φ expresses the anisotropy of the convection. The momentum equation (1) contains the gradient of the turbulent pressure p_t as well as a second term on the left hand side that arises because horizontal motions, in spherical coordinates, can transfer momentum in the radial direction. (In isotropic turbulence $\Phi=3$, and this term vanishes.) The terms involving p_t and Φ come from the Reynolds stress tensor. In the more general case in which the mean and turbulent velocity fields are not as simple as has been assumed here, these terms are more complex and sometimes the concept of an "eddy viscosity" is introduced, appearing in both the mean and the fluctuating equations.

In the mean equations p_t, Φ, and F_c are the quantities that must be determined from the equations for the fluctuations. To solve these a model of the convective turbulence is required. The Boussinesq equations for the fluctuations (Gough 1969,1977a) are:

$$\frac{\partial u_i}{\partial t} + (u_k\partial_k u_i - \overline{u_k\partial_k u_i}) - \frac{\partial \ln(r^2\bar{\rho})}{\partial t}u_i = -\frac{1}{\bar{\rho}}\partial_i p' + \frac{g\bar{\delta}}{\bar{T}}T'\delta_{i3} \tag{7}$$

$$\frac{\partial T'}{\partial t} + (u_k\partial_k T' - \overline{u_k\partial_k T'}) + \left\{ (\bar{C}_{pT} - \bar{\delta})\frac{\partial \ln\bar{T}}{\partial t} - \frac{\bar{\delta}_T\bar{p}\,\bar{\delta}}{\bar{C}_p\bar{\rho}\bar{T}}\frac{\partial \ln\bar{p}}{\partial t} \right\} T' = \beta w - \frac{1}{\bar{C}_p\bar{\rho}}\partial_k F'_k \tag{8}$$

where $\partial_1=\partial/\partial x_1$, $\partial_2=\partial/\partial x_2$, $\partial_3=\partial/\partial x_3$. These equations were obtained by subtracting the mean equations (1) and (5) from the full equations of motion. In equation (8), $C_{pT}=(\partial \ln C_p/\partial \ln T)_p$, $\delta_T=(\partial \ln\delta/\partial \ln T)_p$, and

$$\beta = -\frac{\partial \bar{T}}{\partial r} + \frac{\bar{\delta}}{\bar{C}_p\bar{\rho}}\frac{\partial \bar{p}}{\partial r} \tag{9}$$

is called the "superadiabatic temperature gradient". In the Boussinesq approximation the continuity equation is

$$\partial_k u_k = 0 \tag{10}$$

and the equation of state is

$$\rho'/\bar{\rho} = \delta\, T'/\bar{T}. \tag{11}$$

Both the momentum equation (7) and the energy equation (8) contain terms that depend on the time derivatives of the mean quantities (the last term on the l.h.s. of equation (7) is a consequence of the mean continuity equation, $\partial V/\partial r = -\partial \ln(r^2\bar{\rho})/\partial t$).

A few observations should be made about these equations. Convection can only arise when the superadiabatic gradient β is positive, and in a star β is not a given quantity but a calculated one. (In the static case it is essentially determined by the requirement $\partial_k(\bar{F}_k+F_{ck})=0$. The buoyant acceleration is represented by the last term in equation (7), which has been obtained with the help of equation (11). Viscous terms, which would normally occur in equation (7), even in the Boussinesq approximation, have been neglected because the molecular viscosity in stellar matter is extremely small. The divergence of the fluctuating radiative flux on the r.h.s. of equation (8) represents a loss of thermal energy from a convective eddy, which diminishes the temperature fluctuation.

The nonlinear terms in equations (7) and (8), sometimes called the "turbulent drag" and "turbulent conductivity", respectively, make the system extremely difficult to solve. In stellar convection zones the Rayleigh number, which represents the ratio of buoyant work to energy dissipated by viscous and radiative damping, is large compared with the critical value needed for convective instability. The result is that even perturbations of small length scale can grow, so many modes must be considered. In order to make progress in problems of stellar structure and dynamics, highly simplfied convection models have been adopted. Most commonly used is mixing-length theory, which attempts to describe only the largest-scale modes, that are presumably among the most effective at transporting heat and momentum.

III. MIXING-LENGTH THEORY

The mixing-length theory can be derived essentially from dimensional analysis, but it is helpful to note that it can also be obtained from the Boussinesq equations (7) and (8) using either of two quite different pictures (Gough 1977a). For a static mean atmosphere these equations may be written

$$\frac{\partial w}{\partial t} + (u_k\partial_k w - \overline{u_k\partial_k w}) = \frac{g\bar{\delta}}{\bar{T}}T' \tag{12}$$

$$\frac{\partial T'}{\partial t} + (u_k\partial_k T' - \overline{u_k\partial_k T'}) = \beta w - \frac{1}{\bar{c}_p\bar{\rho}}\partial_k F'_k \tag{13}$$

[In writing equation (12) the gradient of the pressure fluctuation has been neglected. This will be discussed below.]

In the first picture it is supposed that these are the equations of motion of convective eddies that very quickly achieve steady motion. Then $\partial w/\partial t$ and $\partial T'/\partial t$ can be neglected and the nonlinear interaction terms provide dissipation that balances the driving terms on the right hand sides. One estimates these interactions as

$$u_k \partial_k w - \overline{u_k \partial_k w} \simeq w^2/(\ell/2) \tag{14}$$

$$u_k \partial_k T' - \overline{u_k \partial_k T'} \simeq wT'/(\ell/2), \tag{15}$$

where ℓ is a typical length scale of the problem, the "mixing length". The radiative dissipation term in equation (13) is sometimes neglected, but it may not be small in stellar atmospheres and can be approximated in a simple way and included (Böhm-Vitense 1958). Neglecting this term for simplicity, one finds from equation (6)

$$F_c = \overline{C}_p \overline{\rho} (g\overline{\delta}/\overline{T})^{\frac{1}{2}} \ell^2 \beta^{\frac{3}{2}} , \tag{16}$$

the simplest mixing-length expression for the convective heat flux.

In the second picture it is assumed that the motion is not steady, but that an eddy is accelerated as long as it exists. The nonlinear terms are taken to be responsible for the creation and destruction of eddies, but are neglected in the equations of motion. The mixing length enters because now the average lifetime of an eddy is approximated as ℓ/w, and the time derivatives are estimated as

$$\frac{\partial w}{\partial t} \simeq \frac{w}{\ell/w} \simeq \frac{w^2}{\ell}, \qquad \frac{\partial T'}{\partial t} \simeq \frac{T'w}{\ell}.$$

When the fluctuation interactions are neglected, the solution of equations (12) and (13) is the same as in the first picture.

Although the two pictures give the same result in the static case, it was shown by Gough (1977a) that this is not true at all when the mean atmosphere varies in time. The first picture was generalized to the time dependent case by Unno (1967), the second by Gough (1965,1977a). In the latter paper Gough made a detailed comparison of his treatment and Unno's; we shall briefly summarize this comparison and then show how other treatments are related to these two. The present review may be considered a summary and an updating of Gough's (1977a) discussion.

Whichever picture is used, one needs to know how to determine the unknown parameter ℓ, which is an estimate for both the size of an eddy and its mean free path. Many attempts have been made to determine ℓ by comparing properties of stellar models with those of stars. This question will not be considered in this review; a good discussion of the calibration problem was given by Gough and Weiss (1976).

IV. TIME DEPENDENT THEORIES

a. Gough's treatment

The theory developed by Gough (1977a) is quite different from all the others, which have essential features in common. Therefore Gough's work will be discussed first. The essential feature of it is that the nonlinear advection terms in equations (7) and (8) are neglected, as in the second picture described in the previous section. The nonlinearities are taken into account by a detailed discussion of the creation and annihilation of eddies.

Let us rewrite the equations as

$$\frac{\partial u_i}{\partial t} - A(t)u_i + \frac{1}{\bar{\rho}} \partial_i p' = \frac{g\bar{\delta}}{\bar{T}} T'\delta_{i3} \tag{17}$$

$$\frac{\partial T'}{\partial t} + B(t)T' - \beta w = - \frac{1}{\bar{\rho}\bar{C}_p} \partial_k F_k' \tag{18}$$

where the terms that depend on the time derivatives of the mean quantities have been abbreviated as A(t) and B(t).

These linear equations have separable solutions and, so long as the term $\partial_k F_k'$ has a suitably simple form, the spatial form of the eigenfunctions can be described by the horizontal and vertical wavenumbers k_h and k_v. The latter fixes the vertical scale of an eddy, and Gough takes $k_v = \pi/\ell$. The parameter Φ defining the shape of an eddy turns out to be

$$\Phi = 1 + k_v^2/k_h^2. \tag{19}$$

The term involving grad p' in equation (17) can be eliminated by writing an equation for the vorticity which turns out not to involve p'. The net effect of this is that the vertical acceleration term is multiplied by the quantity Φ. This is discussed by Gough, and it is sufficient to mention here that although the term in grad p' is properly included in the fluctuating momentum equation, and in compressible convection contributes to the mechanical work, in the Boussinesq approximation it does no work but simply increases the effective inertia of the vertical flow, since it drives the horizontal motions at the expense of the vertical ones. When the nonlinear terms are retained in the equations of motion one cannot expect the spatial form of the eddies to be so simply described, but nonetheless Unno (1967), Gough (1977a, in describing Unno's theory), and Unno et al. (1979) all treat the fluctuating pressure gradient in a manner similar to that sketched here. Most of the other authors simply ignore it.

The spatial structure having been established, the radial component of equation (17) and equation (18) constitute a set of two ordinary differential equations for w and T' that can be solved if A(t) and B(t) are known, as soon as initial conditions are provided. It is also necessary to specify how the wave numbers (or eddy shape) vary with the motion, and this may be done by assuming that the eddy dimensions distort with the local coordinates. In general A(t) and B(t) may not be known, but in linear pulsation theory the mean quantities, including A(t) and B(t), vary sinusoidally in time, so equations (17) and (18) can readily be integrated to find the motion of an eddy from the time it is born as a local perturbation until it is destroyed. In Gough's theory the lifetime is not simply proportional to the mixing length; rather, the latter is the essential parameter in a probability distribution that determines the rate of destruction of eddies. This probability plays an important role in determining the energy and momentum fluxes (F_c and p_t) to which the motion gives rise. The buoyant elements arise from an instability, and so it is natural to take the birth rate to be proportional to the growth rate of the instability; a statistical distribution of initial conditions is assumed.

In the end, Gough is able to integrate the equations of motion analytically and uses the solutions to compute flux integrals that yield F_c and p_t, which then are substituted into the mean equations. He has emphasized that the turbulent pressure might be important for the excitation or damping of pulsations, even in Boussinesq models, in which it is implicit that $p_t \ll p$. The reason is that a pulsating star works as a kind of heat engine, and it is the component of the pressure that is out of phase with the specific volume that determines the work the engine does. Since the gas pressure tends to have only a small out-of-phase component, a small amount of turbulent pressure could be significant, if it were out of phase. The inclusion of the turbulent pressure and its gradient in the mean equations has the effect of raising by one the order of the set of equations (1)-(5), and has been found to complicate considerably the problem of solving them numerically.

Gough's theory was applied to a one-zone model (Gough 1967) and was used in a study of linearized radial pulsations by Baker and Gough (1979). They found that models at the red edge of the instability strip are stabilized by convection [in agreement with results from nonlinear models by Deupree (1977a,b,c) and Stellingwerf (1984a,b,c)], though the coolest models are again neutral. Gough's theory, like Unno's (§ IV.b.), is a local one, in the sense that convective velocities, fluxes, etc. at a given point are determined solely by the conditions existing at that point. Gough (1977a) showed that inclusion of the turbulent pressure in the mean equations leads, in a local theory, to singularities at the boundaries of the convection zone, and in order to avoid these the turbulent pressure gradient was not computed in a fully consistent way by Baker and Gough (1979). These authors also showed that spatial oscillations in the thermal variables deep in the convection zone are a consequence of the local treatment of convection.

In summary, Gough's treatment of convection has the defect that it omits the continuous damping effects of the interactions with the small-scale motions, which must be expected to limit both the velocity and the temperature fluctuation of an eddy. On the other hand, this theory gives much attention to calculating the acceleration and the energy and momentum fluxes, and this may be very important, because it is known that the damping or driving of pulsations may depend critically on the phases of the fluxes relative to the mean (pulsational) motion. It should be noted that this theory has only been developed for linear radial pulsations.

b. Unno's Treatment

The approach of Unno (1967) is very different from Gough's, being based on the first picture described in § III. The main feature of the model is the approximation embodied in equations (14) and (15) for the advection terms in the fluctuation equations. There are secondary assumptions that simplify the treatment but are not essential to the model: i) The terms we call A(t) and B(t) are neglected, ii) In eliminating the fluctuating pressure gradient it is assumed that $k_v^2 = k_h^2$, and iii) In the mean equations the turbulent pressure is neglected.

A problem arises in describing the change of the mixing length, in its role of eddy size, as the disturbance moves. It is usually set to some small multiple of one of the local scale heights, which we may call H. This defines the dimension of the eddies as they are created, but as the ambient medium distorts, presumably the eddy distorts with it, and hence one might expect the dimension to change with the local Lagrangian length scale. Unno resolves this by choosing the latter for eddies that have lifetimes greater than the pulsation period, and H for eddies having shorter lifetimes. This choice, and the discontinuity it implies, are criticized by Gough (1977a), who also points out that another question is raised: in Unno's picture one does not discuss the creation and destruction of convective elements, and that simplifies the model considerably, but on this particular point it is necessary to bring in some ideas from the other picture.

Unno's nonlinear equations cannot be integrated analytically, and so the equations for w and T' are solved numerically, along with the mean equations (1)-(5). Since the fluctuation equations are local ones, the net effect is just to add two first-order equations (in time) to the system. Unno (1967) used this model to study a very simple one-zone system, and found potentially destabilizing as well as stabilizing effects of convection on the pulsation.

The model was applied to linear radial oscillations of Cepheids by Gonczi and Osaki (1980), who found the reddest models to be only neutrally stable. Later Gonczi (1982a,b) put in the interaction with the mean motion by means of an eddy viscosity term in the momentum equation, and then found Cepheid models to be stable at the red edge of the instability strip.

Although Unno's model treats the nonlinear interactions crudely, and omits other effects that might be important, it has the advantage of producing very simple equations that appear to have interesting consequences. Variations of it have been proposed (Unno 1977, Unno et al. 1979) and it is rather surprising that it has not been more throughly explored and applied.

c. Castor's treatment

The equations we have been discussing describe the motion of convective elements of only a single length scale. The nonlinear terms in the equations couple these modes to others both larger and smaller, but attention is focused on those of larger wavenumber that, through the "turbulent cascade", eventually couple to length scales so small that they can be damped by molecular viscosity. In treatments like Unno's the nonlinear interactions are replaced by damping terms that are meant to describe the loss of energy by the large-scale modes to smaller ones farther down the turbulent cascade.

A number of authors have felt that these nonlinear terms should be replaced by simple expressions describing a diffusion of the turbulent elements in coordinate space, instead of, or in addition to, the local damping picture outlined in the previous section. These authors also work, not with the equations governing the motion of a single representative eddy, but rather with the closely related ones describing the behavior of second-order correlations of the fluctuating quantities.

The first worker to do this for the time-dependent case appears to have been Castor (1968), in a rather widely circulated but unpublished manuscript. Castor postulated a distribution function for turbulent elements, and wrote a conservation equation for this function. [This is closely related to Spiegel's (1963) approach.] It is then easy to develop moment equations, the simplest being those for the second-order correlations. These of course involve the third-order correlations because of the nonlinearities, and the truncation is carried out by approximating the third-order correlations as simple functions of the second-order ones.

In the present context it is convenient instead to carry out this program starting from the equations of motion (17) and (18); the result is the same. Assuming that the "eddy" or "cell" approximation can be used even though the equations are now nonlinear, one can derive from these equations the following:

$$\frac{\partial w}{\partial t} + (u_k \partial_k w - \overline{u_k \partial_k w}) - A(t) w = \frac{g\overline{\delta}}{\Phi \overline{T}} T' \tag{20}$$

$$\frac{\partial T'}{\partial t} + (u_k \partial_k T' - \overline{u_k \partial_k T'}) + B(t) T' - \beta w = - \frac{Kk^2}{\overline{\rho}\,\overline{C}_p} T'. \tag{21}$$

To get equation (20) we have absorbed a factor Φ^{-1} into the definition of A(t) and ignored a like factor in the second term, since it will be crudely approximated in a later step. In equation (21) a radiative conductivity K has been introduced [for details, see Gough (1977a) and Unno et al. (1979)].

Moment equations are obtained by multiplying equations (20) and (21) by w and T' and averaging. After two of the resulting equations are combined we obtain

$$\frac{\partial}{\partial t}\overline{(w^2/2)} + \partial_k \overline{(u_k w^2/2)} - A(t)\overline{w^2} = \frac{g\overline{\delta}}{\Phi \overline{T}} \overline{wT'} \tag{22}$$

$$\frac{\partial}{\partial t}\overline{(T'^2/2)} + \partial_k \overline{(u_k T'^2/2)} + B(t)\overline{T'^2} - \beta\, \overline{wT'} = -\frac{Kk^2}{\overline{\rho}\,\overline{C}_p} \overline{T'^2} \tag{23}$$

$$\frac{\partial}{\partial t}\overline{wT'} + \partial_k (\overline{u_k w T'}) - [A(t)-B(t)]\,\overline{wT'} - \frac{g\overline{\delta}}{\Phi \overline{T}} \overline{T'^2} - \beta\, \overline{w^2} + \frac{Kk^2}{\overline{\rho}\,\overline{C}_p} \overline{wT'} = 0 \tag{24}$$

where equation (10) has been used.

The essential feature of Castor's model is to set

$$\overline{u_k w^2} = -\Lambda_w \overline{w^2}^{1/2} \frac{d\overline{w^2}}{dz} , \tag{25}$$

with similar expressions for the corresponding terms in equations (23) and (24). Λ_w is a diffusion length and $\overline{w^2}^{1/2}$ has the role of a characteristic velocity; their product is a diffusivity, the gradient of which is neglected, so that one writes

$$\partial_k (\overline{u_k w^2}) = -\Lambda_w \overline{w^2}^{1/2} \frac{d^2\overline{w^2}}{dz^2} ,$$

and similarly for the other equations. Usually it is assumed that the three diffusion lengths are all equal, though this is not necessary.

The reason for the assumption (25) is usually not made clear. It can perhaps be looked at in the following way. Separating the kinetic energy into mean and fluctuating parts, we have $\overline{u_k w^2} = \overline{u_k (w^2)'}$. Nothing is known about $(w^2)'$ at the level of the second-order moments, but at least to the extent that it arises from advection, it seems not unreasonable to take it to be proportional to the gradient of $\overline{w^2}$. In a sense this is in a mixing-length spirit, somewhat like the frequently used estimate $T' = \beta\ell$.

Another consideration is the analogy between the Reynolds stress tensor $\sigma_{ik} = -\overline{u_i u_k}$ and the viscous stress tensor τ_{ik}. (The Reynolds tensor is actually $\overline{\rho u_i u_k}$, but the distinction is not important in the Boussinesq approximation.) It is shown in fluid mechanics texts that the term which appears in the kinetic energy equation is $u_i \partial_k \tau_{ik} = \partial_k (u_i \tau_{ik}) - \tau_{ik} \partial_k u_i$, where the first term is the divergence of the viscous energy flux and the second is negative definite and represents the loss of kinetic energy by viscous dissipation. When similar reasoning is applied to $u_i \partial_k \sigma_{ik}$, the mixing-length estimate of the dissipation term is $\overline{w^2}^{3/2}/\ell$, as in the model applied by Unno. If it is now supposed that the convective eddies transport energy and momentum in much the same way that gas molecules do, then the diffusion approximation, applicable if the mean free paths are short compared with the scale lengths of mean quantities, postulates that the flux of a quantity is proportional to the gradient of its mean value. Thus one sets $u_i \sigma_{ik} \simeq -\chi\, d\overline{w^2}/dz$, where the diffusivity χ has the dimensions of length times velocity. The result is

$$\partial_k(\overline{u_i u_i u_k}) = -\Lambda_w \overline{w^2}^{1/2} \frac{d^2\overline{w^2}}{dz^2} + C_w \frac{\overline{w^2}^{3/2}}{\ell} \tag{26}$$

where C_w is a dimensionless parameter. Thus the model can include both diffusion and dissipation. Castor uses a different argument to obtain the dissipation, but his result is the same as equation (26) with $C_w = 1$ and $\Lambda_w = \ell$. The fluctuating pressure gradient does not appear in his equations, nor do A(t) and B(t), but essentially they are the same as equations (22)-(24), with the advection terms replaced according to equation (26) and analogous ones for the other two equations.

Castor applied his time-dependent theory to nonlinear radial pulsations of a single RR Lyrae model. He found the convection zone to be thin and to have little effect on the pulsations. His results might well have been more interesting had he been able to investigate a series of models with a variety of parameters.

d. Stellingwerf's treatment

Castor's manuscript provided the inspiration for the work of Stellingwerf (1982a), who has produced a large number of pulsating-star models with convection (Stellingwerf 1982b; 1984a,b,c). Though his derivation is different from that of Castor or that of the previous section, his equations can readily be obtained from equations (22)-(24). Setting $\Lambda_w = \ell$, $\Phi = C_w = 1$, and combining equations (22) and (26), we find

$$\frac{\partial \overline{w^2}}{\partial t} - \ell\,\overline{w^2}^{1/2}\frac{d^2\overline{w^2}}{dz^2} + \frac{\overline{w^2}^{3/2}}{\ell} - A(t)\overline{w^2} - \frac{g\bar{\delta}}{\bar{T}}\overline{wT'} = 0. \qquad (27)$$

Stellingwerf's treatment now diverges from Castor's, in that equations (23) and (24) do not appear. Instead it is argued that the temperature fluctuation should grow about as fast as the velocity, so one sets

$$T' = \text{const} \times \overline{w^2}^{1/2}, \qquad (28)$$

where the constant of proportionality depends only on mean quantities. The correlation $\overline{wT'}$, which appears in the convective heat flux [equation (6)] as well as in equation (27), is evaluated according to

$$\left.\begin{array}{ll} \overline{wT'} = (\overline{w^2})^{1/2}\,T' & \beta > 0 \\ \overline{wT'} = 0 & \beta \le 0 \end{array}\right] \qquad (29)$$

Theories like Castor's or Stellingwerf's are nonlocal in the sense that the fluctuating quantities at a point can depend on conditions at neighboring points; in particular, the velocities can be nonzero even in regions that are convectively stable ($\beta < 0$), provided they are close enough to an unstable zone. This is called overshooting, and has been the subject of many investigations. Conditions (29) imply that, though elements can overshoot with finite velocities into stable regions, the correlation $\overline{wT'}$, and thus the heat flux, vanish in such regions.

As a practical matter the nonlocal nature of equation (27) is helpful, because the singularities at the edge of the convection zone do not appear, nor do the oscillations deep in the zone (§ IV.a.). (Additional boundary conditions must be supplied, but present no problem.) Hence Stellingwerf is able to include the turbulent pressure gradient in equation (1), though he omits the second term on the left-hand side and adds an "eddy-viscosity" term in an *ad hoc* way.

Stellingwerf (1982b; 1984a,b,c) applied his formulation to models of nonlinear radial pulsations of RR Lyrae stars. He found that convection stabilizes the reddest models, in agreement with previous work of Deupree (1977a,b,c), and also that there are significant effects of convection even in some of the hotter RR Lyrae models. On the whole his convection zones have lower convective velocities than those of the usual mixing-length theory, and the structure of the zones is rather different. Stellingwerf is able to investigate low-amplitude pulsations, providing a comparison with linear models, as well as the effects of turbulent pressure, which is found to be destabilizing in some circumstances and stabilizing in others. Unfortunately a detailed discussion of these interesting pulsating-star models is beyond the scope of the present brief review.

In the following section I shall again refer to Stellingwerf's theoretical treatment and present some criticisms of it. Though the treatment can and should be improved, it must be said that these are the most extensive calculations of time-dependent mixing-length models available, and will surely be a valuable starting point for anyone who wants to study a different formulation.

e. Kuhfuss' treatment

The Stellingwerf formulation was recently reconsidered by Kuhfuss (1986a) who derived similar equations by a rather different method, employing diffusion-type arguments throughout. Diffusion of particles of different species was included. Like Stellingwerf Kuhfuss concentrates on an equation for the kinetic energy, quite similar to equation (27), and does not write a separate equation for T' or $\overline{T'^2}$. There are two important differences: (1) Instead of equation (28) the temperature fluctuation is obtained by an argument which, in our terminology, boils down to

$$T' \simeq \beta \ell, \tag{30}$$

and (2) The convective heat flux $\overline{wT'}$ does not vanish in a stable region where $\overline{w^2}$ is still finite due to diffusion. Instead, since T'<0, by equation (30), the flux is also negative in an overshooting region. [This is rather more like other simple treatments of overshooting, e.g., that of Shaviv and Salpeter (1973).]

The effects of these differences are uncertain, since Kuhfuss has not produced pulsating-star models. He has shown, however, that the differences are potentially significant (Kuhfuss 1986b) and, on the basis of the behavior of the solutions in a static mean atmosphere, has criticized the Stellingwerf theory on two points. First, Stellingwerf's equation of motion requires a nonzero initial velocity, since according to equations (27) and (28) there will otherwise be no acceleration. Kuhfuss shows that the motion can be rather sensitive to the initial conditions, at least in the static case, but Stellingwerf has not discussed them. Second, in a region that in the past was convective, but has meanwhile become stable, the convection in Stellingwerf's formulation cannot die away in a finite time. The convective heat flux, according to equation (29), will vanish, but the motion decays very slowly. Since in Stellingwerf's nonlinear models the convective zone sweeps through the atmosphere leaving stable zones behind it, this effect may have led to an overestimate of the size of the overshooting regions.

One troubling aspect of both the Stellingwerf and Kuhfuss treatments is the fact that, though much attention is devoted to the kinetic-energy equation, very little is given to that for the temperature correlation, $\overline{T'^2}$. Since T' is just as important as w in the expression for the heat flux, which is one of the chief products of the theory, this could be an important shortcoming. Consider the case of linear pulsations having a sinusoidal time dependence. It is well known that, in driving or damping the heat engine, the phase of the flux variation relative to the variation of the temperature gradient is very important, and the phase of F_C depends on that of $\overline{wT'}$. In Stellingwerf's treatment the phase of T' is always that of w; in Kuhfuss' it is that of β. The actual solution to the thermal energy equation is likely to be quite different. Suppose, for example, that the dominant term in equation (8) were βw--then T' would be π/2 out of phase with βw. Such considerations suggests that if one chooses to work with equation (22), equations (23) and (24) should be taken seriously.

f. Xiong's treatment

Using a procedure very much like the one outlined above (§IV.c.), Xiong Da-run (1977, 1979) independently arrived at a formulation very like Castor's. His equations appear to be equivalent to equations (22)-(24), with the estimates given by equation (26) and its analogs for the nonlinear terms. The diffusion terms were added by Xiong (1979) and used in studies of convective overshooting in nonvariable stars (Xiong 1981, 1985). Terms describing mixing of species were included. In the application to variable stars (Xiong 1980) only the dissipation terms were included and the turbulent pressure gradient in equation (1) was neglected in order to avoid the instabilities we have mentioned. The treatment is thus a local one.

Radial linear pulsations of both Cepheids and RR Lyrae stars were studied by Xiong (1980), and he also found that the pulsational growth rates become negative at the cool edge of the instability strip. Thus his conclusions agree with Stellingwerf's, but a detailed comparison of the results is not available. It should be possible to apply a treatment like Xiong's, including diffusion and turbulent pressure, to both linear and nonlinear radial pulsations, but apparently this has not been done.

g. Unno's later work

Unno (1977) reconsidered his earlier treatment, and gave a detailed discussion of the pulsational work integrals, for both radial and nonradial pulsation. The main new aspect was an improved treatment of the variation of the mixing length during pulsation.

Later (Unno et al. 1979) the theory was formulated along the lines of Xiong's local theory. In contrast to Unno's (1967) earlier paper, the effect of the time-varying mean field is specifically included in the fluctuation equations [which are similar to the local version of equations (22)-(24)], and the shape of the eddies is taken account of in a more general way. The formulation is specifically intended for application to nonradial pulsations. I am not aware of any applications to date.

h. Deupree's treatment

Although this brief survey is mainly concerned with a comparison of attempts to generalize mixing-length theory, the completely different approach of Deupree (1977a,b,c) must be mentioned. Deupree made a two-dimensional numerical integration of the full equations, which include both and pulsation and convection. The treatment is both compressible and nonlinear, and is similar to what is called "large-eddy simulation" in meteorology. Such a formulation cannot describe eddies that are smaller than the size fixed by the numerical mesh, so the interactions with smaller length scales ("sub-grid-scale turbulence") are modeled by a time-varying eddy viscosity, the coefficient of which is determined at least partly by the need to control the numerics.

In Deupree's models the convective cells have the height of the entire convection zone, quite contrary to the mixing-length picture, and there is a good deal of evidence from other studies of compressible convection to support this [a good guide to the recent literature can be found in the introductory section of Hurlburt et al. (1986)]. In RR Lyrae models Deupree found stability at the red edge of the instability strip, and gave an interpretation of these results.

There are problems in this work, mainly because of the necessarily poor spatial resolution and the rather *ad hoc* treatment of the time-varying eddy viscosity [see, e.g., criticism by Toomre (1982) and the reply in Deupree (1984)]. Nonetheless it seems clear that as larger computers become more generally available, there will be more attention paid to direct numerical simulations. The results of these will help us to assess the value of the cruder but more tractable (and cheaper) methods that have been the main subject of this review, and may give guidance in refining and calibrating them.

V. NONLOCAL THEORIES

Apart from those approaches that include some diffusion, most of the versions of time-dependent mixing-length theory that have been used are local. Apart from computational difficulties (singlarities, instabilities), there are several objections to this. 1. In places like the H-ionization zone, β varies on a scale much smaller than a local pressure scale height, so an eddy having such a dimension is actually driven (or braked) by a range of values of β. 2. At any given height, heat and momentum transport should depend not only on the eddy centered at that height, but also those located a distances up to $\ell/2$ above or below. 3. In a local theory there can be no penetration into stable regions; an eddy abruptly loses its momentum as soon as it reaches the edge of the unstable zone.

Diffusion theories take some account of the second and third problems, but don't help much with the first. It has been argued that some kind of averaging procedure should be carried out, and many attempts to deal with the overshooting problem are based on this kind of idea. Overshooting as such may not be as important for the time-dependent cases as it is for other problems, but the first two difficulties mentioned above surely are important. Spiegel (1963) suggested an approach to a nonlocal theory, based on treating the motion of convective elements like the transport of radiation in an atmosphere, by means of a conservation equation for their distribution function. The method should be useful for time-dependent convection. More recently Gough (1977b) has surveyed nonlocal theories and developed the ideas somewhat, but averaging methods have never actually been applied, so far as I know, to the time-dependent convection problem.

VI. CONCLUSIONS

A number a different theoretical approaches have been surveyed, and most of them are versions of time-dependent mixing-length theory. Each has its own special characteristics, only the most important of which could be described here, but they are closely related.

Future investigations will probably emphasize nonlocal effects. While overshooting may not play a large role in the pulsation problem, turbulent pressure probably does (cf. Stellingwerf 1982b), and it can be included conveniently only in a nonlocal formulation. In other applications (e.g. novae) mixing may be the most important effect and for this the overshooting has to be calculated. The simplest nonlocal theories are those with diffusion. On physical grounds averaging methods may be preferable, and to this end the suggestions of Gough (1977b) might be developed.

Numerical simulations for both the anelastic and fully compressible equations can now be carried out (at least in two dimensions) for simplified static atmospheres having a depth of several scale heights. The convection zones in these atmospheres are quite different from those obtained with mixing-length models and the overshooting, in particular, is not like that predicted by simple averaging models (cf. Hurlburt et al. 1986). Probably it will be necessary to use simplified models for the time-dependent convection problem in stars for some time, but the results of the large-scale simulations should be kept in mind. It would be worthwhile to develop truly non-Boussinesq theories which are nonetheless simple enough to be used in time-varying stellar atmospheres. An example is the approach to the anelastic equations discussed by Marcus et al. (1983).

Even with the theories now available, or relatively simple extensions of them, it would be desirable to construct models such as simplified stellar atmospheres in order better to understand the physical effects that are important in the pulsation-convection coupling.

I would like to thank R. Kuhfuss for helpful conversations. I think it is also appropriate to acknowledge here my great debt to my old friend Rudi Kippenhahn, without whose 60th birthday half of this conference would not have been held. He and I began working on pulsating stars together 27 years ago, and my continuing association with him and with the Max-Planck-Institut für Astrophysik has meant very much to me. I hope that we shall be celebrating many more of his birthdays.

REFERENCES

Baker, N.H., and Gough, D.O. 1979, Astrophys. J., **234**, 232.
Baker, N.H., and Kippenhahn, R. 1965, Astrophys. J., **142**, 868.
Böhm-Vitense, E. 1958, Zs. f. Astrophys., **66**, 487.
Castor, J. 1968, unpublished.
_____. 1971, Astrophys. J., **166**, 109.
Cox, A.N., Starrfield, S.G., Kidman, R.B, and Pesnell, W.D. 1986, Astrophys. J. (submitted).
Christensen-Dalsgaard, J. 1982, Mon. Not. Roy. Astr. Soc. , **199**, 735.
Deupree, R.G. 1977a, Astrophys. J., **211**, 509.
_____. 1977b, Astrophys. J., **214**, 502.
_____. 1977c, Astrophys. J., **215**, 232.
_____. 1984, Astrophys. J., **282**, 274.

Gonczi, G. 1982a, Astron. Astrophys., **110**, 1.
_____: 1982b, in Pulsations in Classical and Cataclysmic Variable Stars, eds. J.P. Cox and C.J. Hansen (Boulder, JILA), p. 206.
Gonczi, G., and Osaki, Y. 1980, Astron. Astrophys., **84**, 304.
Gough, D.O. 1965, Geophys. Fluid Dynamics, Voll. **2**, (Woods Hole, Mass.: Woods Hole Oceanographic Institution), p. 49.
_____. 1967, Astron. J., **72**, 799.
_____. 1969, J. Atmos. Sci., **26**, 448.
_____. 1977a, Astrophys. J., **214**, 196.
_____. 1977b, in Problems of Stellar Convection, eds. E.A. Spiegel and J.-P. Zahn (Berlin: Springer-Verlag) p. 15.
Gough, D.O., and Weiss, N.O. 1976, Mon. Not. Roy. Astr. Soc., **176**, 589.
Hurlburt, N.E., Toomre, J., and Massaguer, J.M. 1986, Astrophys. J. (in press).
Kuhfuss, R. 1986a, Astron. Astrophys., **160**, 116.
_____. 1986b, in Proc. Second Recontre on Nuclear Astrophysics, ed. J. Audouze (Paris) (submitted).
Marcus, P.S., Press, W.H., and Teukolsky, S.A. 1983, Astrophys. J., **267**, 795.
Shaviv, G., and Salpeter, E.E. 1973, Astrophys. J., **184**, 191.
Spiegel, E.A. 1963, Astrophys. J., **138**, 216.
Spiegel, E.A., and Veronis, G. 1960, Astrophys. J., **131**, 442 (correction: **135**, 665).
Toomre, J. 1982 in Pulsations in Classical and Cataclysmic Variable Stars, eds. J.P. Cox and C.J. Hansen (Boulder, JILA), p. 170.
Unno, W. 1967, Publ. Astr. Soc. Japan, **19**, 140.
_____. 1977, in Problems of Stellar Convection, eds. E.A. Spiegel and J.-P. Zahn (Berlin: Springer-Verlag) p. 315.
Unno. W., Osaki, Y., Ando, H., and Shibahashi, H. 1979, Nonradial Oscillations of Stars (Tokyo: Univ. Tokyo Press) §§ 12, 23.
Xiong, Da-run 1977, Acta Astron. Sinica, **18**, 86 [Engl. trans. in Chinese Astronomy, **2**, 118 (1978)].
_____. 1979, Acta Astron. Sinica, **20**, 238 [Engl. trans. in Chinese Astronomy, **4**, 234 (1980)].
_____. 1980, Scientia Sinica, **23**, 1139.
_____. 1981, Scientia Sinica, **24**, 1406.
_____. 1985, Astron. Astrophys., **150**, 133.

Infrared Giants

L.B. Lucy

European Southern Observatory, Karl-Schwarzschild-Str. 2,
D-8046 Garching, Fed. Rep. of Germany

1. Introduction

Towards the end of their lives, many stars become enshrouded in an envelope of self-created dust, thereby becoming faint and wellnigh unobservable optically, but bright and prominent on the infrared sky. The observational indications are that this intrinsically intriguing phenomenon is also associated with a fractionally large cumulative mass loss and thus signals a decisive turning point in a star's evolutionary history. As such, the incorporation of this phenomenon and the concomitant mass loss into stellar evolution codes is evidently highly desirable. Among the important issues whose proper quantitative investigation would then be facilitated are the following: the upper mass limit on the main sequence for ultimate evolution into white dwarfs; the mass function for white dwarfs; the structure of the immediate progenitors of planetary nebulae; and the chemical and particulate enrichment of the interstellar medium. A useful review of these topics in the context of the theory of asymptotic giant branch (AGB) stars has been given by Iben and Renzini (1983).

Hitherto, investigators of stellar evolution with mass loss have simply peeled matter from their models in accordance with the dictates of some formula for Φ, the rate of mass loss. For cool stars, the formula adopted has usually been that of Reimers (1975), which has $\Phi \propto L/MR$. The validity of its application to dust-enshrouded, chemically peculiar stars at the tip of the AGB is doubtful, however, since it is not from such stars that this formula is derived or calibrated. Moreover, since there is no theoretical reason for expecting a simple scaling law to hold, the problem does not reduce to that of determining the appropriate value of the proportionality constant for this type of star.

The preferred approach in computing evolutionary tracks for mass-losing stars is surely to derive the instantaneous mass loss rate from the basic equations governing the structure of the expanding circumstellar envelope. Then, just as our current codes do not rely on spectroscopic analyses for photospheric pressures, future codes would not rely on such analyses for mass loss rates. Pursuing this analogy, we might further hope that, as with stellar atmosphere theory for wind-free stars, the relevant wind theory will prove sufficiently amenable to simplification that the effects of the wind can be taken into account via the star's surface boundary conditions.

For luminous AGB stars, a major obstacle to this purely theoretical approach is uncertainty about the basic mass loss mechanism - in particular, about the relative rôles of pulsation and dust formation. Thus, a plausible case can be made that surface matter is lost because it is propelled outwards by a succession of pulsation-generated shocks, with dust formation then being an inconsequential downstream phenomenon. For the cool carbon stars, however, a case can be made that dust formation followed by the radiative acceleration of the frictionally coupled gas-dust mixture is the basic phenomenon and that pulsation is just coincidentally and unavoidably present, being indeed a common characteristic of cool giants.

Of the above alternatives, the first poses serious technical difficulties - e.g., computing the radiative losses from shocks propagating in and partially dissociating a molecular gas - that must be solved before reliable mass loss rates can be predicted. The second alternative is therefore adopted here simply because progress is then possible and so consequences can be evaluated. Accordingly, in this investigation, the evolution of an AGB star is picked up at the point when it becomes a carbon star, followed into the dust-condensation domain, and continued with modified photospheric boundary conditions derived on the assumption that the star then acquires a steady, dust-driven wind. The evolutionary tracks thus obtained will presumably lead to conflicts with observational data if the mass loss of such stars is in fact fundamentally a pulsation phenomenon.

2. Outline of theory

If the luminosity - core mass relation for AGB stars (Paczynski, 1970; Uus, 1970) is used, many aspects of their evolution can be investigated with a simple stellar envelope code incorporating the mixing-length treatment of surface convection zones. Moreover, this approach is readily extended to the case where the star is assumed to have developed a dust-driven wind.

When calculating an AGB star and associated wind in this way, we have the freedom to specify its luminosity L as well as the total mass below its "surface" at r = R. However, we cannot specify the radius R (which must be defined) nor the values at r = R of the physical variables P, T and v. There are thus four quantities to be determined and thus we must impose four constraints on the equations governing the envelope structure of our mass-losing star. Once these quantities are known, the mass loss rate follows since

$$\Phi = 4\pi r^2 \rho v \qquad (1)$$

is a constant of the flow and so can be evaluated at r = R.

The constraints that determine the solution are the following: Inward integrations of the convective envelope must give

$$M(r) \rightarrow M_c \quad \text{as} \quad r \rightarrow 0 \,, \tag{2}$$

where M_c is the core mass predicted for the specified L by the adopted L-M_c relation. The second constraint, which constitutes the definition of the star's radius (Lucy, 1976), is that

$$\tilde{\tau}\,(R) \quad = \quad \frac{2}{3} \,, \tag{3}$$

where

$$\tilde{\tau}\,(r) \quad = \quad \int_r^{\infty} \kappa\rho \left(\frac{R}{r}\right)^2 dr \,, \tag{4}$$

The third constraint is that

$$T \rightarrow 0 \quad \text{as } r \rightarrow \infty \,, \tag{5}$$

which follows from the assumption of no external radiation field. The final constraint demands that

$$S = S_c \quad \text{at} \quad v = a \,. \tag{6}$$

This condition states that, at the sonic point ($v = a$) in the expanding envelope, the supersaturation ratio S for carbon vapour over graphite must equal the critical value S_c for sudden condensation.

Notice that neither S nor $\tilde{\tau}$ are additional independent variables. The former is a function of state variables and the latter is related algebraically to the temperature via the modified Milne-Eddington formula (Lucy, 1976)

$$T^4 \quad = \quad \frac{1}{2}\, T_*{}^4 \left(2W + \frac{3}{2}\,\tilde{\tau}\right) \,, \tag{7}$$

where W is the dilution factor and T_* is the surface temperature.

Further explanation is perhaps required in respect of condition (6). To a good approximation, the momentum equation for a steady, dust-driven wind may be written as

$$(v^2 - a^2)\,\frac{1}{v}\,\frac{dv}{dr} \quad = \quad g_R - g \,, \tag{8}$$

where g_R is the radiative acceleration. Regularity at the sonic point thus requires that $g_R = g$ when $v = a$. Moreover, the desired solution obeying this condition has $g_R < g$ for $v < a$ and $g_R > g$ for $v > a$ in order that v increase outwards. Now, given the low opacity of cool gaseous matter and the possibly high opacity of a dust-gas mixture, the only way to obtain such a solution is for the sonic point to be in the dust condensation zone. Accordingly, with the further assumption of instantaneous dust condensation when S increases to S_c, this condensation point must be coincident with the sonic point - condition (6). In the subsonic flow matter is then entirely gaseous and $g_R < g$, whereas in the supersonic flow dust formation is complete and $g_R > g$.

Figure 1 illustrates the contemplated structure of mass-losing AGB carbon stars. Notice that the dust-condensation radius R_d is shown as significantly exceeding the radius R. This is a consequence of the need to dilute the photospheric radiation field before the temperature can drop to the value required for dust condensation.

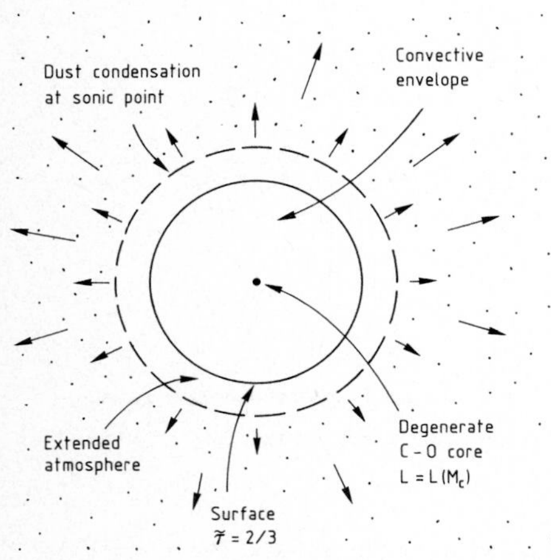

Fig. 1: Sketch of AGB carbon star's structure when in dust-enshrouded, high mass loss rate phase.

3. Simplified theory

Detailed solutions for the wind and atmospheric structure implied by the theory sketched out above were published already ten years ago (Lucy, 1976). Now, in principle, such solutions could be directly matched to solutions generated by a convective envelope code to obtain complete mass-losing AGB models. But since the wind solutions alone require numerous shooting integrations in two iteration loops in order to determine the theory's two eigenvalues and a further such loop must be added to satisfy condition (2), some simplification is desirable, if not mandatory.

The basic simplification adopted, which eliminates one iteration loop, is to constrain the wind's terminal velocity v_∞ to have a pre-assigned value. This is achieved in the following way: we first assume that the absorption coefficient κ of the gas-dust mixture is grey and constant. Then, with the further assumption that

$v^2 \gg a^2$ throughout the supersonic flow, Eq. (8) can be integrated analytically, the result being

$$v^2 = v_\infty^2 \left(1 - \frac{R_d}{r}\right) , \tag{9}$$

with

$$v_\infty^2 = v_{esc}^2 (1 - \delta) , \tag{10}$$

where $\delta = \kappa L / 4\pi GMc$ and v_{esc} is the escape velocity from the dust condensation radius R_d. Thus, for a given trial value of R_d and an assigned value of v_∞, Eq. (10) may be solved for δ and therefore κ. In effect, therefore, a mean value of the opacity of the dust-gas mixture is obtained by demanding that radiative driving accounts for the observed v_∞.

A further simplification - convenient but not essential - is the approximation of the loci along which $S(P,T) = S_c$ by the isotherms $T = T_d(S_c)$. Thus, for a given choice of S_c, dust condensation occurs at temperature T_d, and this must therefore be the temperature given by Eq. (7) at $r = R_d$. In this way, for given trial values of R and R_d, we obtain $\tilde{\tau}_w$, the value of $\tilde{\tau}$ at R_d. But now this same quantity can also be obtained by performing the integration in Eq. (4) using Eqs. (1) and (9) together with the previous assumption that κ is a constant. The result is

$$\tilde{\tau}_w = \frac{16}{15} \frac{\kappa \, \Phi \, R^2}{4\pi \, v_\infty \, R_d^3} , \tag{11}$$

which then allows Φ to be calculated from the values of $\tilde{\tau}_w$ and κ derived above. With Φ thus determined, the density at the sonic point is given by Eq. (1) and this then completes our knowledge of physical conditions at this point since we already know that $T = T_d$, $v^2 = a^2 = kT_d/\mu m_H$, and $r = R_d$.

The assumption of hydrostatic equilibrium becomes a good approximation in the subsonic flow when $v^2 \ll a^2$. The intermediate zone immediately below the sonic point where this assumption is not valid can conveniently be treated as a discontinuity. Conservation of momentum flux then requires the pressure on the subsonic side of the discontinuity to be higher by $\Delta P = \rho a^2$, with the other variables T, $\tilde{\tau}$ and r being unchanged.

With physical conditions at the top of the star's extended atmosphere thus determined, its structure is obtained by inward integration of the equation of hydrostatic equilibrium, with temperature again given by Eq. (7) but now with the further increase in $\tilde{\tau}$ being due to the opacity $\kappa(\rho,T)$ of the dust-free gas. Now this integration must satisfy Eq. (3) which, for a given trial value of R, serves therefore to determine R_d.

With physical conditions at the star's surface $r = R$ thus determined, its interior structure is obtained with a convective envelope code, and this integration must satisfy Eq. (2). If it does not, then R must be changed and R_d recomputed. Thus we have two iteration loops to determine the two eigenvalues R and R_d.

The above discussion describes how the instantaneous structure of a mass-losing AGB star is obtained. When such a model has been calculated its mass loss rate Φ is known, and so its new mass after a time step Δt can be computed. Moreover, since the rate of growth of M_c is determined by L, its new core mass and therefore L can also be computed. Thus the evolutionary track of the mass-losing AGB star is readily calculated.

4. An example

To illustrate the simplified theory described above, the evolution of an AGB star of mass 3 $M_\odot$ is computed following its supposed conversion into a carbon star at M_{bol} = -4. Throughout the subsequent mass-losing phase, its wind's terminal velocity is taken to be v_∞ = 15.8 km s^{-1}, which is the average of the observed values for 25 carbon stars in the Knapp-Morris (1985) survey of CO emission from circumstellar envelopes. In addition, we take T_d = 1660 K corresponding to S_c = 2 and mixing-length ratio α = 0.5. Finally, the atmospheric opacities are those of Sharp (unpublished), which include the important polyatomic molecules HCN and C_2H_2.

The evolutionary track followed by this model is plotted in Fig. 2. Because effective temperature is not well defined for stars with extended photospheres, the abscissa is log T_*, where T_* is the temperature at $\tilde{\tau}$ = 2/3. This reduces to T_{eff} when the atsmophere is not extended.

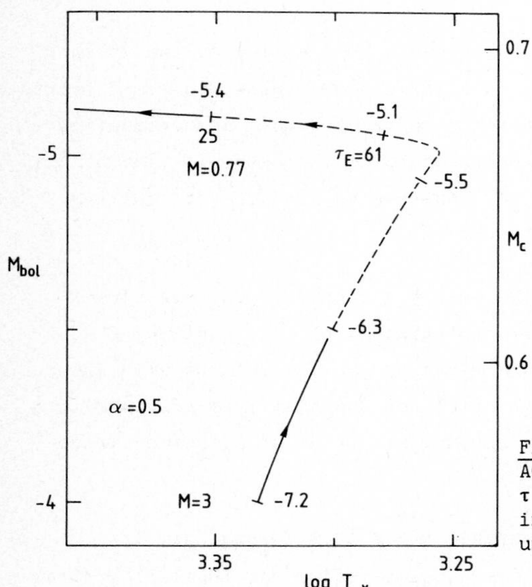

Fig. 2: Evolutionary track for mass-losing AGB carbon star. Dashed portion has $\tau > 2/3$ for dust-free matter. Mass M is in solar units and the time scale τ_E in units of 10^3 yrs.

The initial model in Fig. 2 has a weak and optically thin wind with $\Phi = -7.2$ dex. However, as L increases and T_* drops, both Φ and the optical thickness of the circumstellar envelope increase. In particular, when $M_{bol} \simeq -4.5$, we have $\Phi = -6.3$ dex and dust-free matter - i.e., $r < R_d$ - is all at optical depth $\tau > 2/3$ from an external observer. In this circumstance, most of the radiation emitted by the star's gaseous photosphere is absorbed and re-emitted by dust in the circumstellar envelope; consequently, the bulk of the star's radiation finally emerges in the infrared. In this dust-enshrouded phase - the dashed part of the evolutionary track - the mass loss rate reaches the maximum value $\Phi = -5.1$ dex and the cumulative mass loss is 2.09 $M_\odot$ - i.e., 70% of its initial mass as a carbon star. These aspects of this track are encouragingly similar to empirical estimates (Knapp and Morris, 1985, and references therein).

Another quantity of interest plotted on Fig. 2 is the e-folding time τ_E for the reduction of $M-M_c$, the remaining envelope mass. This drops to a minimum of 25,000 yrs but increases thereafter as the blueward evolution weakens and then terminates the dust-driven wind. This time scale would need to be reduced by a factor ~10 if this wind alone is to account for the formation of a planetary nebula - central star configuration. It seems likely, therefore, that the theory presented above does not provide a complete description of the mass loss history of stars at the tip of the AGB.

More details of the theory and further examples of evolutionary tracks will be published elsewhere.

References

Iben, I. Jr., Renzini, A.: 1983, Ann. Rev. Astron. Astrophys. **21**, 271.

Knapp, G.R., Morris, M.: 1985, Astrophys. J. **292**, 640.

Lucy, L.B.: 1976, Astrophys. J. **205**, 482.

Paczynski, B.: 1970, Acta Astron. **20**, 47.

Reimers, D.: 1975, Mem. Soc. R. Sci. Liège, 6e Ser. 8, 369.

Uus, U.: 1970, Nauch. Inform. Acad. Nauk. **17**, 3.

On the Solutions for Contact Binary Systems *

A. Weigert
Hamburger Sternwarte, D-2050 Hamburg 80, Fed. Rep. of Germany

Although the pioneering paper of Lucy (1968) appeared already 18 years ago, the theory of contact binaries is by far not yet in a satisfactory state. This concerns in particular an aspect with which I will not deal here at all, namely the inability to reproduce theoretically the observed systems (light curves, period-colour relation). This might indicate that some basic physical effects are still missing. But even within the frame of the presently used description, rather little is known of the solutions and their stability. This is due to the fact, that one has to deal with a whole variety of unknown parameters. On the other hand, it is cumbersome to obtain numerical solutions. Consequently, only rather few of them were available in the literature until recently. The present review is based on three papers by Kähler, Matraka and Weigert (1986a,b,c = papers I,II,III) who recently presented and discussed a large number of new numerical results.

1. The Problem

Consider a contact binary system of given angular momentum J and total mass $M = M_1 + M_2 = M_1(1 + q)$. Following the usual theoretical description of such systems we have first the wellknown differential equations of stellar structure applied to both components:

Diff. eqs. of stellar structure for component 1
Diff. eqs. of stellar structure for component 2 (1)

Since the components can exchange mass and luminosity in their common envelope, we need 2 additional contact conditions (see,e.g., Hazlehurst, Refsdal, 198o). The first is the requirement of equal potentials ϕ_i (i = 1, 2) at the two surfaces. Defining $F := \phi_1 - \phi_2$, we write the condition for given J and M as

*) Dedicated to my friends R. Kippenhahn (6o) and H.U. Schmidt (6o) who were the first astronomers I met more than 36 years ago.

$$F(M_1, M_2, R_1, R_2) = 0 \quad (2)$$

The second condition describes the amount Λ of luminosity which is exchanged between the components if they have slightly different specific entropies S_1, S_2 in their envelopes. For this transfer equation we use an approximation derived by Hazlehurst and Meyer-Hofmeister (1973)

$$\Lambda = K \cdot d^m \cdot (S_1 - S_2)^n \quad (3)$$

where d is the depth of contact which can also be expressed in terms of the R_i, M_i and J. The S_i are the specific entropies in the two envelopes. Since K, m and n are not well known from physical arguments, they have to be varied as free parameters.

The differential system (1), (2), (3) is solved for given M, J and chemical composition by a very efficient numerical method described in paper I. It is a Henyey-type procedure for the system as a whole using a Henyey matrix H of the structure

$$H = \begin{pmatrix} (H1) & & \\ & (cont) & \\ & & (H2) \end{pmatrix} \quad (4)$$

The usual Henyey matrices for the single components (H1) and (H2), are here connected by the matrix (cont) derived from the contact conditions. The determinant $|H|$ bears also very valuable information on the stability of the solution and on certain properties of a linear series of such solutions.

For the thermal stability problem, it was shown (paper I) in particular that

$$\text{sign}(|H|) = -1^k \quad (5)$$

where k is the number of unstable modes (real or complex). Obviously, sign $(|H|) = -1$ is sufficient for instability. A case with sign $(|H|) = +1$ needs an additional eigenvalue analysis in order to distinguish between stability (k = o) and instability (k = 2, 4 ...). When $|H| = 0$, there is marginal stability with an eigenvalue $\sigma = 0$, and a linear series of solutions has a critical point (e.g. a turning point).

In the dynamical stability problem, one considers adiabatic perturbations with constant M and J. Using the function F in (2), the condition for dynamical stability is found to be

$$\left(\frac{dF}{dM_1} \right)_{S,M,J} > 0 \quad . \quad (6)$$

2. Equilibrium Solutions

The time independent equilibrium solutions are discussed in detail in paper III. The "natural" parameters to be given for a system are the chemical composition (which is supposed to be homogeneous, except if stated otherwise), the total mass M and the angular momentum J. A solution yields then also the mass ratio $q = M_2/M_1$, for which we will often use $\xi = \ln q$ for reasons of symmetry. Many such solutions obtained for the same chemical composition and M = const, but varying J, define the function $q = q(J)$, or correspondingly $\xi = \xi(J)$. We report a few of the results obtained for different parameters K, m, n.

a) Symmetric solutions, $q = 1$ ($\xi = o$): Two identical components can always be in deeper or shallower contact with $\Lambda = o$ (independent of d, K, m, n) since $S_1 = S_2$. The only condition is that the contact is not broken. Such systems are therefore possible for all $J \leq J_{max}$ (cf.Fig.1).

b) Neighbourhood of $q = 1$: For an infinitesimal neighbourhood of $\xi=\Lambda=o$, an analytical treatment is possible by using a linearized expansion in terms of ξ and Λ (except for the very critical transfer equation (3) which has to be used without linearization). One can show (paper III)that the solutions obey

$$(J_{max}^2 - J^2)^m \cdot \xi^{n-1} = C \tag{7}$$

where the constant $C = C(M, K, m, n)$. Obviously, the solutions depend now strongly on the parameters K, m, n. For the different cases, the possible neighbouring solutions are sketched in Fig.1 which shows J^2 over $\xi = \ln q$.

Case C = o: Neighbouring solutions bifurcate from the symmetric ones at $J = J_{max}$ (cf. Fig. 1a).

All following cases assume $C > o$.

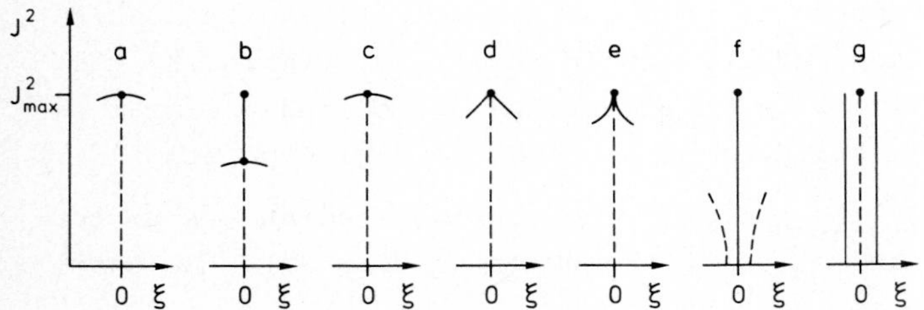

Fig.1. Possible solutions at and around $\xi = o$ for different cases of m,n

Case m ≠ o, n = 1: The neighbouring solutions leave the symmetric ones below J_{max} (Fig. 1b).

Case m ≠ o, n < 1: The neighbouring solutions start at $J = J_{max}$, but with different tangent depending on the value of the parameter $\lambda = (1-n)/m$. $\lambda > 1$ is sketched in Fig.1c, $\lambda = 1$ in Fig. 1d, and $\lambda < 1$ in Fig. 1e.

Case m ≠ o, n > 1: The neighbouring solutions run as shown in Fig.1f. They are completely separated from the symmetric solutions (due to the condition $J^2 > o$) by $\xi_o = C^{1/(m-1)} \cdot J_{max}^{2/\lambda}$.

Case m = o, n ≠ 1: The neighbouring solutions are located on parallel straight lines (Fig. 1g).

Case m = o, n = 1: No neighbouring solutions exist for $C \neq 1$. If $C = 1$, the neighbouring solutions form a 2dimensional continuum for all $\xi \ll 1$ and all $J < J_{max}$.

c) Arbitrary mass ratio: Outside the just treated small neighbourhood of $q = 1$, the solutions have to be obtained numerically. This was done for several values of the parameters K, m, n. A few results are represented in Fig. 2. Close to $q = 1$, some of the cases discussed in Fig. 1 are easily recovered. Similar to the earlier condition $J \leq J_{max}$ for the

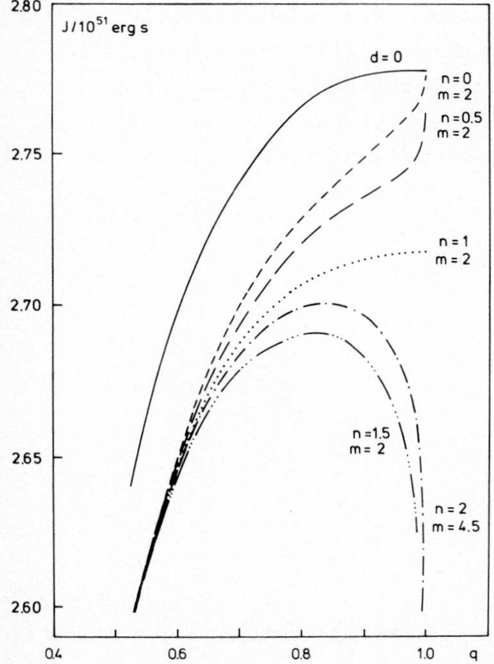

Fig. 2. Solutions outside q = 1.

symmetric solution, there is for any value of q an upper limit of J above which the contact is broken. Correspondingly the curve for the limiting case of zero contact depth, d = o, defines an upper limit for all other cases. In the other curves K is chosen such that all of them have equal d at values of q far from 1. In that range, the curves for different m,n become very similar since they are simply dominated by the orbital properties of the system (e.g. the change of the volumes of the Roche lobes). In the vicinity of q = 1, however, the dominating influence comes from the transfer equation (3) and therefore the solutions differ strongly for different m,n. Since one cannot hope to cover all values of the parameters in detail, we will in the following concentrate on the interesting cases with $m \neq o$, $n > 1$ (for example the combination n = 2, m = 4.5).

For fixed values K, $m \neq o$, $n > 1$ and M = const, the solutions are once more illustrated in Fig.3 giving $\xi = \xi(J)$. This representation can be considered as a linear series of solutions with J varying as the parameter. The series consists of 5 branches. For given J, one has either 1 ($J_1 < J \leq J_{max}$), or 5 solutions ($J < J_1$). With the strict chemical homogeneity treated here, there is full symmetry with respect to $\xi = o$ (q = 1), and 2 of the solutions are trivially obtained from 2 others by a mere exchange of the components. But this is certainly different if one assumes a slight inhomogeneity of the composition, say, in the interior of the primary. Such a case was also treated numerically (paper III). The result is that no symmetry is left (of course, also no symmetric solution at q = 1), and all 5 solutions for given J have become essentially different. The linear series has turning points at $J = J_1$, where 2 branches merge. In these critical points one finds $|H| = o$. This means also that the "local uniqueness" of the solution is violated (compare paper I).

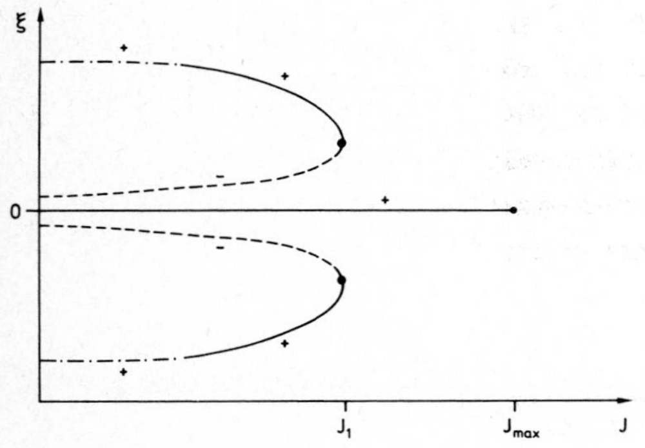

Fig.3. Solutions for $m \neq o$, $n > 1$.

The next step is then to calculate the solutions for different total masses M (but the same K,m,n). In this way one can obtain an overview over the possible solutions for this case (compare, e.g.Fig.4 in paper III). Instead of going into detail here, we rather turn to the question whether these equilibrium solutions are stable against small perturbations.

3. Stability of Equilibrium Solutions

Let us first discuss the thermal stability. We concentrate again on the case $m \neq o$, $n > 1$ and look once more at the solutions plotted in Fig.3. The sign of $|H|$ is indicated along the different branches. There are only two critical points with $|H| = o$ and consequently $\sigma = o$ (at $J = J_1$). The solutions on the dashed branches have $|H| < o$ and are therefore unstable with an odd number of real modes. The positive branches had to be analyzed additionally. In the case indicated in Fig.3 they were found to contain stable solutions (on the solid lines) as well as solutions with 2 unstable conjugate complex modes (dot-dashed lines). The border of this complex mode instability turned out to depend very much on the total mass M. The border can even move over the the turning points on the negative branches such that there are no stable solutions left for $\xi \neq o$.

An overview over the many complicated possibilities can be obtained from Fig.4 which is constructed for m = 4.5, n = 2. In this diagram M_1

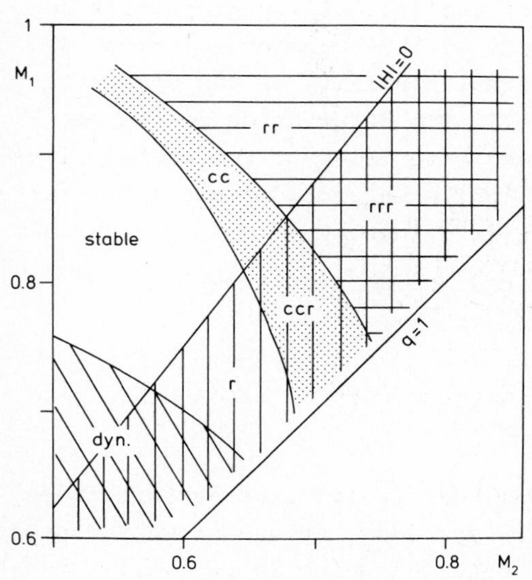

Fig.4. Regions of different stability properties

is plotted over M_2, such that $M = M_1 + M_2$ = const on a diagonal along which the mass ratio q (or ξ) varies. The curves in Fig.3 (which is for M = const) can now be projected on this diagonal and we obtain the borders between the region of different stability properties. Aside the region of stability one sees areas in which quite different modes are unstable. Each label "r" indicates one unstable mode, "c" one unstable complex mode. The sector between the lines q = 1 and $|H|$= o contains the solutions on the dashed branches in Fig.3, with the turning point at $J = J_1$ being projected on the line $|H|$ = o.

The detailed eigenvalue analysis which was necessary in order to single out the different unstable modes revealed some interesting aspects. As an instructive example (which will be dealt with again in the next section) we consider the change of the eigenvalues when passing from the "stable" region in Fig.4 through that labeled "cc" into "rr" along a vertical line (M_2 = const) to increasing M_1. It has turned out that all eigenvalues (real or complex) remain almost stationary, except for the two highest ones on the negative real axis. During the transition, they merge, split up into a conjugate complex pair which crosses the imaginary axis when the border to the region "cc" is reached. Later they merge again on the real positive axis and split up into a real pair in the region "rr". A very peculiar point is the "four instability corner" where the line $|H|$ = o intersects the border between "cc" and "rr". At that intersection, there are 3 coinciding eigenvalues σ = o which split up in different ways depending on the sector into which the solution shifts.

Tests have shown that the details of the instability properties depend appreciably on the parameters K,m,n of the transfer equation (3). This is true in particular for the thermal stability of the symmetric solutions q = 1; as can be seen in Fig.1. The bifurcation points are critical points of the linear series and have $|H|$ = o; the dashed lines ($|H|$ < o) indicate unstable solutions, while the solid lines ($|H|$ > o) have turned out to contain only stable solutions.

Finally, there is an area of dynamical instability shown in the lower left corner of Fig.4.

4. Thermal Relaxation Oscillations (TRO)

We now turn to time-dependent solutions. Of special interest are those known as TRO which can occur in certain systems. Let us ask what happens if we slightly perturb an equilibrium solution which is thermally unstable via complex modes (situated in the region "cc" of Fig.4). Obviously, the perturbed system will oscillate periodically around the equilibrium state (corresponding to Jm $\sigma \neq o$) with increasing amplitude (since Re $\sigma > o$). The solution approaches finally a TRO with a period of the order $1o^7$ years. This represents a limit cycle, i.e. a nonlinear solution which is strictly periodic as long as one neglects effects of chemical evolution.

The period and the amplitude of the TRO depend on the properties of the "coordinate" equilibrium model the perturbation of which yields the TRO. In all cases treated in paper II, the period of the (non-linear) TRO turned out to be $2\pi/$ Im σ, where σ is the eigenvalue determined from the (linear) thermal stability analysis of the coordinate equilibrium model.

The results for the amplitudes were also surprising. Consider again the earlier discussed sequence of equilibrium models with increasing M_1 and M_2 = const, reaching from the stable region into that labelled "cc" in Fig.4. With increasing M_1 along this sequence, the perturbation yielded TRO of increasing amplitude. This can be seen in Fig.5 showing the extrema of $\dot{M}_1$ as a measure for the amplitude of the TRO over M_1 of the coordinate equilibrium model. In a typical phase diagram, say, $\dot{M}_1$ over M_1, the TRO of larger amplitudes do not encircle the coordinate equilibrium solution. (The same is true for the motion in the H-R diagram.)

The most surprising result was that the amplitude of the TRO did not go to zero for the coordinate equilibrium model situated on the

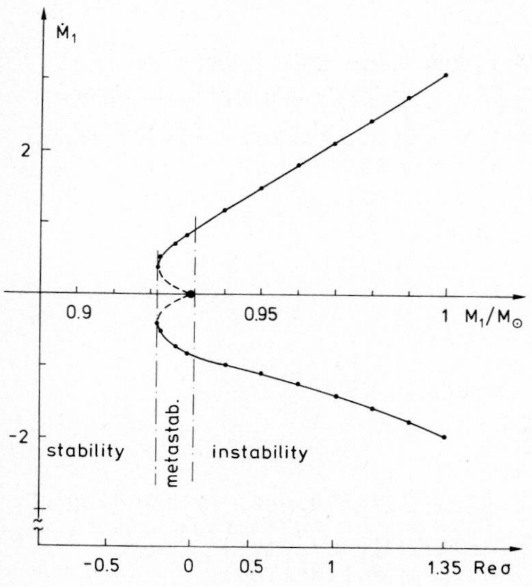

Fig.5. Amplitudes of relaxation oscillations depending on M_1 of the unperturbed model

border to the stable region (Fig.5). Instead, the TRO persisted to exist for some range on the stable part of the sequence where they could be obtained by a finite amplitude perturbation of the (linearly) stable equilibrium models. This means that these models are metastable (stable against infinitesimal perturbations, but unstable against finite perturbations). Therefore, Fig.4 should be supplemented by a narrow strip of metastability separating the unstable region "cc" from the really stable configurations.

The dashed lines in Fig.5 indicate the amplitude of the perturbation which was necessary in order to drive the stable models into TRO. It acts like a watershed between linear and nonlinear behaviour. This is schematically indicated in Figure 6 which shows some twodimensional phase diagram with the watershed (dashed) beyond which perturbed solutions go to the TRO (heavy solid line) while perturbations inside the watershed fall back to the equilibrium state (dot). Fig.5 shows that the extension of the region inside the watershed goes to zero for the model at the border to linear instability.

5. Limit Cycles and Bifurcations

In terms used in nonlinear dynamics the TRO are a limit cycle which is stable since small deviations (say, by the unavoidable numerical errors) are damped off rapidly. It could be shown (paper II) that the above described "watershed" forms another limit cycle, however, an unstable one. This means that any infinitesimal deviation from the cycle makes the solution run away either to the stable limit cycle, or to the equilibrium state (cf. also Fig.6).

In our simple sequence of models reaching from the stable to the unstable region "cc" in Fig.4, there are 4 types of steady solutions: fixed points and limit cycles, both stable and unstable. This is shown

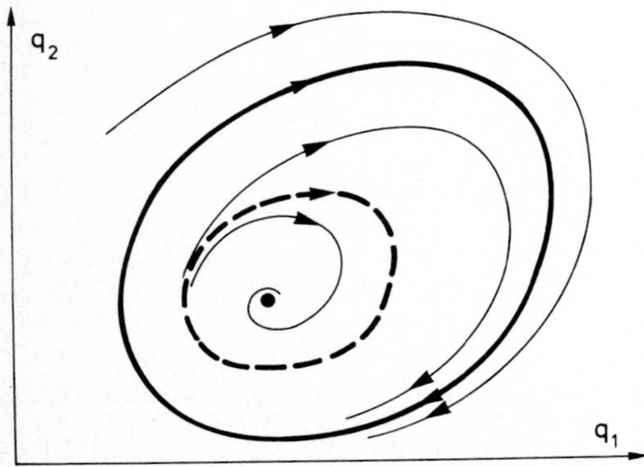

Fig.6. Solutions approaching either the equilibrium (dot) or the stable limit cycle (heavy solid line)

in Fig.7 where the "amplitude" r of the different solutions is plotted over a parameter Γ varying along the sequence (say, a measure for M_1). In this representation, the steady solutions form a linear series with several branches. The stable fixed point acts as an attractor for the whole region 1 of attraction, while the stable limit cycles act as attractor for solutions starting in region 2 of attraction. At the border between stable and unstable fixed points ($\Gamma = 1$) there is a bifurcation where the unstable limit cycle leaves the branch of fixed points. The branches of unstable and stable limit cycles finally merge in a turning point ($\Gamma = o$). The branch of unstable limit cycles merges into the branch of fixed points in a bifurcation at the border between stable and unstable fixed points ($\Gamma = 1$), i.e. where a pair of complex eigenvalues of the linear stability problem crosses the imaginary axis. Such type of bifurcation (which gave rise to the metastability in the region between bifurcation point and turning point) is known as a subcritical bifurcation.

The classical Hopf bifurcation with a stable limit cycle leaving the fixed point is described by a very simple set of equations. Of course, one can not hope to have a similar analytical description for the complicated solutions of contact binaries in full detail. But the topologically equivalent structure of the linear series as shown in Fig.7 can well be reproduced already by a very simple analytical example in 2 dimensions. Using polar coordinates and introducing the control parameter Γ we consider the differential equations

$$\frac{dr}{dt} = (\Gamma - 1)\, r + 2\, r^3 - r^5 \; ; \quad \frac{d\theta}{dt} = \omega\, (\Gamma) \qquad (7)$$

Obviously, one has the steady ($dr/dt = o$) solution $r = o$, $\theta = \omega t$, which represents a fixed point. It is stable ($dr/dt < o$) against in-

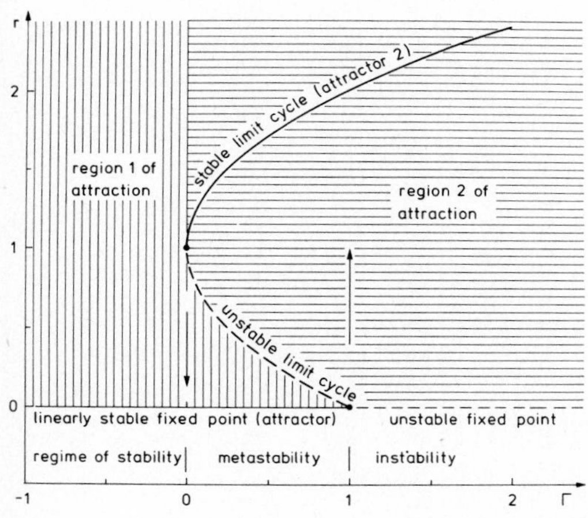

Fig.7. Linear series containing fixed points and limit cycles

finitesimally small perturbations for $\Gamma < 1$, and unstable $(dr/dt > o)$ for $\Gamma > 1$, as the linearized equations (7) show. Then there is a steady solution $r = (1 + \sqrt{\Gamma})^{1/2}$ (giving again $dr/dt = o$), $\theta = \omega t$ for $\Gamma > o$, i.e. a limit cycle which can be shown to be stable. And finally one has an unstable limit cycle given by the steady solution $r = (1 - \sqrt{\Gamma})^{1/2}$, $\theta = \omega t$ for $o < \Gamma < 1$. Clearly, these steady solutions are just those plotted in Fig.7.

References

Hazlehurst,J., Meyer-Hofmeister,E.: 1973, Astron.Astrophys. 24, 379

Hazlehurst,J., Refsdal,S.: 198o, Astron.Astrophys. 84, 2oo

Kähler,H., Matraka,B., Weigert,A.: 1986 a, Astron.Astrophys. 151, 317

Kähler,H., Matraka,B., Weigert,A.: 1986 b, Astron.Astrophys. 161, 296

Kähler,H., Matraka,B., Weigert,A.: 1986 c, Astron.Astrophys. in press

Lucy,L.B.: 1968, Astrophys.J. 151, 1123

White Dwarfs in Close Binaries

H. Ritter

Universitäts-Sternwarte München, Scheinerstr. 1,
D-8000 München 80, Fed. Rep. of Germany

We explore to what extent the available observational data of white dwarfs in binary systems agree with theoretical predictions about the white dwarfs produced during a binary evolution. We show that if the observational data are carefully examined and, as in the case of cataclysmic binaries, proper allowance for the most important selection effects is made, the observations are consistent with theoretical predictions at least as far as the mean mass of the white dwarfs is concerned.

1. INTRODUCTION

It is common knowledge that about two out of every three stars are members of either a binary or a multiple system and that about 95% of all stars end their life as a white dwarf. From this it follows directly that binary systems that contain a white dwarf must be extremely abundant. However, the importance of white dwarfs in binaries for stellar astrophysics does not mainly derive from their high abundance. (As a matter of fact, due to the intrinsic faintness of white dwarfs, most of these binaries escape detection.) Rather these systems are important because they provide key information for our understanding of the structure and evolution of white dwarfs and of the evolution and interaction of close binary systems. Only if a white dwarf is a member of a binary can its dynamical mass and, in favourable cases, its radius be determined accurately so that the results may be compared with Chandrasekhar's (1931a,b; 1935) classical prediction or with a more refined theory (e.g. Hamada and Salpeter, 1961).

Theoretical understanding of how degenerate stars form from unevolved ones had to wait for the development of an efficient way to compute stellar evolution numerically. Twenty years ago, Kippenhahn, Kohl and Weigert (1967) showed in what is now a classical paper how white dwarfs can be formed in close binary systems via Roche-lobe overflow. This paper must still be considered as one of the great achievements in binary evolution theory, all the more so since essentially all the basic principles that govern the formation of a white dwarf in a close binary were outlined and have stood the passage of time without major changes. The impact of that paper on binary evolution theory was enormous. In its aftermath numerous studies devoted to the formation of white dwarfs in close binaries have been published (for a review up to 1978 see Webbink, 1979; for later work e.g. Trimble, 1983; Webbink, 1985).

The main emphasis of this contribution is, however, not to review binary evolution theory but rather to explore to what extent its quantitative predictions regarding white dwarfs match with observational data. For this, we begin with a brief overview of the observational data in section 2 and of the theoretical predictions in section 3. A comparison of theory with observations in section 4 leads us to a more detailed investigation in section 5 of the masses of the white dwarfs in cataclysmic binaries. Finally, our conclusions are presented in section 6.

2. OBSERVATIONS OF WHITE DWARFS IN BINARIES

Although the number of binary systems that contain a white dwarf component must be very large indeed, only a small fraction of them has provided us with quantiative information about their degenerate components. These systems may be conveniently subdivided into the following three classes:

2.1. Detached, long-period systems

Many of these systems with an orbital period of a few decades and longer are known. However, so far only the four systems listed in Table 1 have allowed a dynamical determination of their white dwarf's mass and (in only two cases) of its radius. In fact, the first two systems in Table 1 are still the only ones where the mass and the radius of the white dwarf can be determined independent of each other with some accuracy. This means that the empirical verification of the theoretically predicted mass-radius relation (Chandrasekhar, 1931a,b; 1935; Hamada and Salpeter, 1961) rests on only two white dwarfs: 40 Eri B and α CMa B = Sirius B.

2.2. Detached, short-period systems

These systems consist of a white dwarf or a white dwarf precursor (sdOB star) and a low-mass ($\lesssim 1\ M_\odot$) companion (usually unevolved) and have orbital periods in the range $0\overset{d}{.}1 \lesssim P \lesssim$ few days. Only a small

Table 1

Orbital parameters of the four detached, long-period binary systems that contain a white dwarf of known mass. P is the orbital period, e the eccentricity of the orbit, M_{WD} and R_{WD}, respectively, the mass and the radius of the white dwarf and M_{comp} the mass of the companion(s).

System	P(yr)	e	$M_{WD}/M_{\odot}$	$R_{WD}/R_{\odot}$	$M_{comp.}/M_{\odot}$	Ref.
σ^2 Eri BC	252.1	0.410	0.43±0.02	0.013±0.003	0.16±0.01	1,2
α CMa AB	50.09	0.592	1.053±0.028	0.0073±0.0012	2.143±0.056	3,4
α CMi AB	40.65	0.40	0.63		1.74	5
Stein 2051	23	0.3	0.50±0.05 0.72±0.08		0.23 (0.19+0.17)	6

References, Table 1

1 Heintz (1974)
2 Wegner (1980)
3 Gatewood and Gatewood (1978)
4 Holberg et al. (1951)
5 Strand (1951)
6 Strand (1977)

Table 2

Orbital parameters of 10 precataclysmic binaries with a white dwarf of known mass. P is the orbital period, M_{WD} and M_2, respectively, the mass of the white dwarf and of the companion. A detailed discussion of the adopted masses is given in Ritter (1986b). A + in the column CPN? indicates that the object is also the central star of a planetary nebula.

System	CPN?	P(days)	$M_{WD}/M_{\odot}$	$M_2/M_{\odot}$
V651 Mon	+	15.991	0.32-0.45	1.8±0.3
FF Aqr		9.20755	0.5	2.0
Feige 24		4.2319	0.46-1.24	0.25±0.05
BE UMa		2.291168	0.6±0.1	0.2-0.6
UX CVn		0.573703	0.39±0.05	0.48±0.05
V471 Tau		0.521183	0.71±0.01	0.73±0.03
UU Sge	+	0.465069	1.05	0.6
LSS 2018	+	0.35713	0.7±0.1	0.31±0.04
GK Vir		0.34433081	0.42-1.08	0.14-0.67
AA Dor		0.26515398	0.2-0.3	0.043±0.006

number of these systems (at present 22) which are considered to be the immediate precursors of cataclysmic binaries (e.g. Ritter, 1983, 1986b; Webbink, 1986) are known. A compilation of their orbital parameters is given by Ritter (1984; 1987). A detailed discussion of their masses and other properties of evolutionary interest has recently been given by Ritter (1986b). In what follows we shall refer to these systems as precataclysmic binaries (PCBs).

In the context of this paper the masses of the white dwarf components are of primary interest. At present, individual masses can be derived for only 10 of the 22 systems. The main orbital parameters of these 10 systems are listed in Table 2. Since in most cases the accuracy of an individual mass determination is rather low, we restrict ourselves to the discussion of the mean white dwarf mass of this sample. From Table 2 we find a mean value of $\langle M_{WD} \rangle = (0.62 \pm 0.08) M_\odot$ and a standard deviation of $\sigma_M = 0.24\ M_\odot$.

2.3. Semi-detached, short-period systems

These systems consist of a white dwarf and an unevolved low-mass companion that fills its critical Roche volume. Systems of this type are usually referred to as cataclysmic binaries (CBs). For a review see e.g. Warner (1976). At present some 10^3 cataclysmic variables (CVs) are known. However, a more or less accurate orbital period is known for only about 115 of them. A compilation of their orbital parameters is given in Ritter (1984, 1987). The masses of the white dwarfs in CBs have been a matter of dispute ever since their first systematic investigation by Warner (1973). This is because Warner's result that the mean white dwarf mass is $\langle M_{WD} \rangle = (1.2 \pm 0.2) M_\odot$ was hard to reconcile with basic predictions of binary evolution theory (see below, section 3.2). Since the masses of CBs are notoriously difficult to determine, much of the dispute has focused on the question as to whether the various approaches for determining the masses yield reliable results. In order to escape this dispute as much as possible we restrict ourselves (at least for the moment) to the discussion of the most reliable masses available. These derive from the few (12) double-lined spectroscopic CBs AE Aqr, V363 Aur, Z Cam, AC Cnc, BV Cen, Z Cha, SS Cyg, EM Cyg, U Gem, AH Her, RU Peg and IP Peg. Their masses and other parameters are listed in Table 3. Among the above 12 objects there are, however, only 6 that are also eclipsing (V363 Aur, AC Cnc, Z Cha, EM Cyg, U Gem and IP Peg). Only for these can the masses be determined directly. For the others a solution for the masses requires one additional assumption. The one most frequently used is that the low-mass companion is on the main sequence and obeys a given mass-radius relation. The above sample of 12 systems (i.e. sample No. 1 in Table 4) yields a mean white dwarf mass of $\langle M_{WD} \rangle = (0.90 \pm 0.06) M_\odot$ with a standard deviation of $\sigma_M = 0.22\ M_\odot$. The fact that the corresponding values for the 6 eclipsing and the 6 non-eclipsing systems are, respectively, $\langle M_{WD} \rangle = (0.80 \pm 0.09) M_\odot$, $\sigma_M = 0.23\ M_\odot$ and $\langle M_{WD} \rangle = (1.01 \pm 0.07) M_\odot$, $\sigma_M = 0.16\ M_\odot$ might indicate that using a standard mass-radius relation for the secondary of the non-eclipsing systems results in a systematic overestimate of the masses. However, we will show later (section 5.3) that this difference can be explained in a different way. For the moment we assume that the best available observational data of CBs yield a mean white dwarf mass of $\langle M_{WD} \rangle \approx 0.9\ M_\odot$.

3. THEORETICAL PREDICTIONS

Before we proceed in comparing the observational results summarized in section 2 with theoretical predictions we give here a brief outline of the basic theoretical principles on which these predictions are based. For this we begin with a short review of the formation of isolated white dwarfs.

3.1. The formation of isolated white dwarfs

Isolated white dwarfs are the former degenerate carbon oxygen cores of stars that have lost their envelope via a stellar wind and/or the ejection of a planetary nebula during their evolution on the asymptotic giant branch (e.g. Schoenberner, 1979, 1983). Because stellar winds are not arbitrarily efficient in removing the stellar envelope and thus in limiting the growth of the degenerate core, only stars having an initial mass M_i less than a certain upper limit M_u (to be specified) will yield a degenerate core of mass M_f that is below the Chandrasekhar mass M_{CH}. Since on the asymptotic giant branch the wind loss rate $\dot{M}_W$ is most sensitive to the stellar luminosity L (Reimers, 1975; Kudritzki and Reimers, 1978) and since, in turn, L is essentially determined only by the mass of the degenerate core M_c, i.e. by the core mass luminosity relation (e.g. Paczyński, 1970; Kippenhahn, 1981) and at the same time by the rate $\dot{M}_c$ with which the core grows, one expects a tight correlation between M_f and M_i. Due to our incomplete theoretical understanding of the wind loss mechanism on the asymptotic giant branch we cannot determine theoretically a reliable M_i-M_f-relation. On the other hand, observations of white dwarfs in open clusters (e.g. Koester and Reimers, 1981, 1982) have allowed the determination of an empirical M_i-M_f-relation (Weidemann and Koester, 1983; Weidemann, 1984a). We note two important properties of this relation. First, white dwarfs with $M_f = M_{CH}$ originate from stars with $M_i = M_u \approx 8\text{-}10\ M_\odot$. Second, the M_i-M_f relation is rather flat for small M_i such that the white dwarfs that are descendants of stars with $M_i \lesssim 3\ M_\odot$ fall into a narrow mass range $M_f \approx 0.5\text{-}0.65\ M_\odot$. Because the initial mass function of main sequence stars is steep, i.e. $dN(M_i)/dM_i \sim M_i^{-2.35}$ (Salpeter, 1955; Miller and Scalo, 1979) and because the M_i-M_f-relation is so flat for small M_i, the vast majority of single stars with masses $M_i < M_u$ end as white dwarfs in the narrow mass range $0.5\ M_\odot \lesssim M_f \lesssim 0.65\ M_\odot$. In fact, the observed mean mass of single white dwarfs is $\langle M_{SWD} \rangle \approx 0.6\ M_\odot$ with a standard deviation of $\sigma_M = 0.1\ M_\odot$ (Koester, Schulz and Weidemann, 1979; Koester and Weidemann, 1980; Schönberner, 1981; Weidemann and Koester, 1984), in accordance with the above prediction.

3.2. The formation of white dwarfs in close binaries

The numerous ways through which white dwarfs can emerge from the evolution of a close binary have been reviewed in great detail by Webbink (1979, 1985). Here we are not so much concerned with the various types of bianry interaction but rather with their outcome in terms of the mass distribution of the resulting white dwarfs. By narrowing the scope of our discussion to the masses we are left with just three different modes of white dwarf production:

3.2.1. Formation of a CO white dwarf via case C mass transfer

In case C evolution (e.g. Lauterborn, 1970) the primary loses its envelope via Roche-lobe overflow while it is on the asymptotic giant branch (after central helium burning but before carbon ignition). Therefore, our above discussion of the formation of isolated white dwarfs is of relevance here. This is because the M_i-M_f-relation yields the upper limit for the mass M_f of a white dwarf that descends from a primary of inital mass M_i from Roche-lobe overflow while on the asymptotic giant branch. The corresponding lower limit $M_{f,l}$, on the other hand, is the core mass at the base of the asymptotic giant branch. The function $M_{f,l}(M_i)$ is not only flatter than $M_f(M_i)$ but, more importantly, for any $M_i < M_u$ we have $M_{f,l} < M_f$. Thus the mass of a white dwarf formed in case C evolution is $M_{f,l}(M_i) < M_{CO\text{-}WD} < M_f(M_i)$, i.e. it is always smaller than that which a single star with the same M_i would yield. Because Roche-lobe overflow strips the primary on a time scale that is much shorter than the nuclear time scale, the value of M_u for which $M_f(M_u) = M_{CH}$ can be higher than the corresponding value of $8\text{-}10\ M_\odot$ for single stars. However, it cannot be significantly higher than $\sim 10\ M_\odot$ because for higher masses the core after central helium burning will not be degenerate and thus the evolution to larger radii and binary interaction will be suppressed. Therefore, since the distribution of the initial primary masses is just as steep as the initial mass function of single stars, the mass spectrum of the resulting CO-white dwarfs will be similar to that of single white dwarfs but shifted to slightly smaller masses. We would thus expect a mean value $\langle M_{CO-WD} \rangle \lesssim 0.6\ M_\odot$.

3.2.2. Formation of He-white dwarfs via low-mass case B mass transfer
In low-mass case B evolution the primary with a mass $M_i \lesssim 2.5\ M_\odot$ undergoes Roche-lobe overflow on the first giant branch, i.e. after central hydrogen burning but before helium ignition (Kippenhahn, Kohl and Weigert, 1967). In this case the upper limit of the white dwarf mass is the helium flash mass $M_{He-Fl} \approx 0.45\ M_\odot$, the lower one the mass of the degenerate core after central hydrogen burning, which is of the order of 0.15-0.20 $M_\odot$. Therefore we would expect that the mean mass of He-white dwarfs is somewhere between these two limits. In any case it is significantly smaller than the $\approx 0.6\ M_\odot$ of the CO-white dwarfs. In addition to this we conclude from Fig. 2.2.6 in Webbink (1985) that low-mass case B evolutions occur at least as frequently as case C evolutions. However, it is still an open question as to whether close binaries containing a He-white dwarf are formed with the same frequency since little is known about the outcome of the common-envelope evolution (e.g. Meyer and Meyer-Hofmeister, 1979) that is initiated through mass transfer from the giant. It may well be possible that a substantial fraction of these evolutions end in coalescence, i.e. in a single star if the degenerate component has only a small mass (Livio and Soker, 1984).

3.2.3. Formation of a white dwarf via high-mass case B mass transfer
In this case the primary with a mass $M_i \gtrsim 2.5\ M_\odot$ undergoes Roche-lobe overflow after central hydrogen burning but before helium ignition and becomes a helium star (Kippenhahn and Weigert, 1967). The evolution following the formation of the helium star depends mainly on its mass and on whether a second mass transfer phase (a so-called case BB evolution) occurs. Case BB has been studied by De Grève and De Loore (1976, 1977), Delgado and Thomas (1980), Habets (1986), and with regard to the formation of massive white dwarfs in CBs by Law and Ritter (1983). The main result of the latter investigation is that case BB is unlikely to contribute a significant fraction of the (massive) white dwarfs in CBs. This is because this type of evolution requires a rather massive primary and has thus a correspondingly low

frequency of occurrence. The same conclusion may also be drawn from Figure 2.2.6 in Webbink (1985).

From the above arguments we can thus conclude that the mass spectrum of the CO-white dwarfs in close binaries is probably very similar to that of single white dwarfs but shifted to slightly smaller masses. The mean mass of the white dwarfs in close binaries will, therefore, mainly depend on the as yet unknown contribution of the He-white dwarfs. If it is small the mean mass is expected to be close to 0.6 $M_\odot$. If, on the other hand, it is significant the resulting mean mass is correspondingly smaller.

4. CONFRONTATION OF THEORETICAL AND OBSERVATIONAL RESULTS

We are now in a position to make a meaningful comparison between theoretical expectations and observational results. We concentrate on the verification of the mass-radius relation and on whether theory predicts correctly the mean white dwarf masses in different types of binaries.

4.1. The mass radius relation

As we have already noted above (section 2.1) the observational verification of the mass-radius relation rests on just two white dwarfs, 40 Eri B and Sirius B (see Table 1), although Chandrasekhar's (1931a,b, 1935) theoretical prediction has stood now for 50 years. How do these observations compare with theory? The data for Sirius B clearly rule out the $R_{WD} \sim M_{WD}^{-1/3}$ mass-radius relation defended by Eddington (1935). However, the observational errors in M and R of both white dwarfs are still too large to allow clues about their internal composition to be obtained from the more refined composition-dependent mass-radius relations of Hamada and Salpeter (1961). Although there are no doubts that the theoretical predictions are correct, the current state of affairs with only two white dwarfs to test these predictions, is unsatisfactory. Yet the situation is unlikely to improve significantly in the forseeable future.

In this context it is worth re-emphasizing Weidemann's (1984b) point that earlier claims according to which there were 7 in 1958 and that by now there are hundreds of white dwarfs that fall exactly on Chandrasekhar's mass-radius relation (Schatzmann, 1958; Wali, 1982) are completely unsubstantiated.

4.2. The masses of the white dwarfs

4.2.1. Detached, long-period systems

Although attempts have been made to account for the properties of the Sirius system in terms of a close binary evolution (Lauterborn, 1970), the long orbital period (and thus the large semi-major axis) together with the high orbital eccentricity in all the systems in Table 1 suggest that the former primary was never close to filling its critical Roche-lobe and that, therefore, these white dwarfs were formed in the single star mode (for a discussion of the Sirius system see D'Antona, 1982). Consequently the mean mass of these white dwarfs $\langle M_{WD} \rangle \approx 0.7\ M_{\odot}$ must not be compared with predictions of binary evolution but rather with predictions from single star evolution. Although the sample of four white dwarfs is too small to give such a comparison much weight, it is evident that in the light of the theoretical expectations two of the four white dwarfs (40 Eri B and Sirus B) are, to say the least, unusual. A posteriori probabilities are always extremely low. Nevertheless it is a bit embarassing that the most massive white dwarf known is also the nearest one. While Sirus B can at least be accounted for by normal single star evolution (though the formation of a 1.05 $M_{\odot}$ white dwarf must be a very rare event indeed), 40 Eri B with its mass of 0.43 $M_{\odot}$, which lies below the He-flash mass, presents a more serious problem. Explanation of this star requires that a star of initial mass $M_i \gtrsim 0.9\ M_{\odot}$ lost its envelope (mass $M_e \gtrsim 0.47\ M_{\odot}$) on the first giant branch, presumably via a stellar wind. This in turn implies uncomfortably high wind loss rates, i.e. a very efficient mass loss mechanism.

4.2.2. Precataclysmic binaries

The mean of the observed white dwarf masses $\langle M_{WD} \rangle = (0.62 \pm 0.08) M_{\odot}$ agrees reasonably well with theoretical expectations. Unfortunately most of the individual masses (see Table 2) are sufficiently uncertain and the number of systems available is too small to allow definite conclusions about the contribution of He-white dwarfs to be drawn. Nevertheless, the degenerate components of V651 Mon, UX CVn and AA Dor are promising candidates.

4.2.3. Cataclysmic binaries

The best available observational data (see sample No. 1 in Table 4) yield a mean white dwarf mass of $\langle M_{WD} \rangle = (0.90 \pm 0.06) M_{\odot}$. This value is at variance with the theoretical prediction that $\langle M_{WD} \rangle \lesssim 0.6\ M_{\odot}$. The result of 0.9 $M_{\odot}$ is even more puzzling in the light of the fact that the PCBs as the immediate evolutionary precursors of CBs have white dwarf masses that are in line with theory. The fact that the white dwarfs in CBs are uncomfortably massive is not new. It has been known since Warner's (1973) pioneering study and has generated a lively discussion about how this discrepancy might be accounted for. A satisfactory solution to this problem has only very recently been given. It will be discussed in the next section.

5. THE MASSES OF THE WHITE DWARFS IN CATACLYSMIC BINARIES

5.1. Possible solutions to the mass problem

The attempts that have been made in the past to account for the high masses of the white dwarfs in CBs are based mainly on one of the following arguments:

a) The observationally derived masses (see Table 3) are systematically too high. Although systematic errors in deriving masses are probably not negligible, they should be least important for the double-lined spectroscopic CBs (sample No. 1 in Table 4) which yield the high mean mass of 0.9 $M_{\odot}$. Therefore, the high masses are most probably real and not primarily the result of systematic errors.

b) The intrinsic mass spectrum of the white dwarfs could be significantly different from that of single white dwarfs because in binaries white dwarfs can be formed in ways that are not open to single stars. In fact, it follows from our arguments in section 3.2 that a non-negligible fraction of CBs could contain a low-mass He-white dwarf. As a result we would expect a <u>lower</u> rather than a higher mean mass of the white dwarfs in CBs as compared to single white dwarfs.

c) The high masses could be a consequence of persistent accretion, i.e. of the secular evolution of CBs. This hypothesis has attracted much attention in the past (for a review see e.g. Ritter, 1983, 1986a), mainly in the context of the origin of Type I supernovae. However, it is now clear that this hypothesis can be rejected for the vast majority of CBs (Mac Donald, 1984; Ritter, 1983, 1986a).

d) The high masses result from observational selection. As is described in the next paragraph this is probably the solution to the problem.

5.2. Properties of a magnitude-limited sample of CBs

Despite earlier negative results of Livio and Soker (1984), Ritter and Burkert (1985, 1986), Ritter and Özkan (1986) and Ritter (1986c) have recently studied observational selection among CBs in favour of massive white dwarfs. For this they computed the properties of a V-magnitude-limited sample of those CBs that have a bright quasi-stationary accretion disk, i.e. of dwarf novae during an outburst and of nova-like systems of the UX UMa subtype. In these computations it was assumed that the accretion disk is the only relevant source of luminosity. The accretion rate that determines the accretion luminosity was thereby obtained from a simple model of the secular evolution of CBs. The main results of these computations can be summarized as follows:

a) Restriction to a magnitude-limited sample results in an enormous selection effect in favour of massive white dwarfs. The strength of the selection effect is expressed in terms of the selection function

$$S(M_{WD}) = \frac{p_o(M_{WD})}{p_i(M_{WD})} \tag{1}$$

where $p_o(M_{WD})$ is the distribution of the white dwarf masses in the magnitude-limited sample and $p_i(M_{WD})$ the intrinsic mass spectrum of the white dwarfs in CBs. An approximate analytical solution of the selection problem (Ritter, 1986c) for the apparently brightest systems, say $m_V \lesssim 12$ (for dwarf novae in outburst), yields

$$S(M_{WD}) = \text{const.} \left(\frac{M_{WD}}{R_{WD}}\right)^{3/2} \int_{M_2} \dot{M}_{WD}^{1/2}\, 10^{0.6BC}\, dM_2 \tag{2}$$

where R_{WD} is the radius of the white dwarf, $\dot{M}_{WD}$ the accretion rate, BC the bolometric correction of the accretion disk and M_2 the mass of the secondary star. The integral in (2) has to be performed over all possible secondary masses. It turns out that this integral is only a slowly varying function of M_{WD}. Therefore the selection function is dominated by the factor $(M_{WD}/R_{WD})^{3/2}$ which is a steeply increasing function of M_{WD}. Unfortunately, $p_i(M_{WD})$ is not known, hence, we cannot compute $p_o(M_{WD})$. However if we accept that the intrinsic mass spectrum of the CO-white dwarfs in CBs is similar to that of single white dwarfs $p_{SWD}(M)$ (see section 3.2.1), we find that the selection effect is strong enogh to account fully for the high mean white dwarf mass observed.

b) As is immediately clear from Eq. (2), the main reason for the strong selection is the mass-radius relation of white dwarfs. Therefore, we can now turn the argument around and conclude that the observed high masses of the white dwarfs in CBs confirm indirectly the mass-radius relation as derived by Chandrasekhar (1935).

c) The observed white dwarf masses are fully compatible with predictions of binary evolution theory.

d) The fact that selection in favour of massive white dwarfs is so strong must mean that we observe only a minority of CBs and that, therefore, their intrinsic space density must be much higher than has been inferred from observations. In fact, if the local space density of dwarf novae of $\sim 10^{-6} pc^{-3}$ found by Duerbeck (1983) is to be reproduced by our computations, then the corresponding intrinsic space density of CBs must be of the order of $n_{CB} \approx 1\text{-}2\ 10^{-4} pc^{-3}$. Assuming that the birthrate of CBs has been roughly constant over the past 10^{10}yr, we arrive at a local birth rate of $\dot{n}_{CB} \approx 1\text{-}2\ 10^{-14}$ $pc^{-3} yr^{-1}$. This may be compared with the birthrate of single white dwarfs $\dot{n}_{SWD} \approx 10^{-12} pc^{-3} yr^{-1}$ (Koester and Weidemann, 1980; Fleming, Liebert and Green, 1986).

e) The majority of Type I supernovae, having a birthrate of $\sim 10^{-13} pc^{-3} yr^{-1}$, cannot descend from CBs.

f) Our computations predict that the mean white dwarf mass of a magnitude-limited sample of CBs depends systematically on the observable properties of a sample. The most important of these selection trends are:

- The mean white dwarf mass increases with the mean value of the secondaries' masses or with the mean value of the orbital period of the sample: $\partial \langle M_{WD} \rangle / \partial \langle M_2 \rangle > 0$ and $\partial \langle M_{WD} \rangle / \partial \langle P \rangle > 0$.
- In a sample that contains dwarf novae and nova-like systems, the mean mass of the white dwarfs is higher in dwarf novae than in nova-like systems.
- Below the period gap of CBs, i.e. for orbital periods $P \lesssim 2^h$, no UX UMa systems should exist.
- The mean white dwarf mass is a weakly increasing function of the mean orbital inclination. In particular, this means that eclipsing systems should have a slightly higher mean white dwarf mass than the non-eclipsing systems.
- The selection effect becomes weaker, i.e. the mean white dwarf mass smaller with increasing limiting magnitude m_V of the sample, i.e. $\partial \langle M_{WD} \rangle / \partial m_V < 0$.

5.3. Comparison with observations

In order to check whether the above-predicted selection trends can be verified with the available observational data, we have collected the best, i.e. in general the most recent, estimates for the white dwarfs' masses of all dwarf novae and nova-like (UX UMa) systems that are either single-lined or double-lined spectroscopic binaries from the literature. The relevant data for these systems are listed in Table 3. In addition, Table 3 includes the double-lined, nova-like system AE Aqr although it belongs to the DQ Her rather than to the UX UMa subtype. However, AE Aqr enters only samples No. 1 and 2 in Table 4 (see below). Including AE Aqr, Table 3 lists 37 systems. These represent about half of all dwarf novae and UX UMa systems or about 1/3 of all known CBs. Thus the sample should be representative. However, it is not magnitude-limited in the sense of our computations since the degree of completeness decreases strongly with increasing m_V. Comparison of the predicted trends with the data in Table 3 is, nevertheless, meaningful since the aforementioned incompleteness does not change the sign of the trends.

According to the above-predicted selection trends we have drawn various subsamples from Table 3. Their defining characteristics together with their statistical properties are listed in Table 4. With these data we can now check our predictions:

a) The predicted increase of $\langle M_{WD} \rangle$ with $\langle P \rangle$ is confirmed from samples 1-10. These data are also shown in Figure 1. The dashed line representing the mean trend corresponds to the relation

$$\langle M_{WD} \rangle = 0.43\ M_{\odot} + 0.068\ M_{\odot}\ \langle P(h) \rangle \tag{3}$$

This trend is a natural consequence of the fact that CBs are stable against mass transfer (e.g. Ritter, 1976, 1986a) and had been noticed earlier by Shafter (1983). Stability requires that for any given orbital period (or secondary mass) the mass of the associated white dwarf must not be much less than that of the secondary. Thus the dependence of $\langle M_{WD} \rangle$ on $\langle P \rangle$ or $\langle M_2 \rangle$ is not restricted to a magnitude-limited sample. The fact that $\langle M_{WD} \rangle$ in the samples 1, 2 and 7 is so high is thus at least in part a consequence of the fact that double-lined spectroscopic CBs are predominantly long-period systems.

This dependence is probably also the explanation for the difference in $\langle M_{WD} \rangle$ of the eclipsing and non-eclipsing double-lined CBs mentioned in section 2.3. The respective mean orbital periods are $\langle P \rangle_{ecl} = 5\overset{h}{.}3 \pm 1\overset{h}{.}0$ and $\langle P \rangle_{non-ecl} = 8\overset{h}{.}9 \pm 1\overset{h}{.}3$. Using Eq. (3), the difference of $\Delta\langle P \rangle = 3\overset{h}{.}6$ transforms into $\Delta\langle M_{WD} \rangle = 0.24\ M_{\odot}$. Within the errors this is exactly what is observed.

b) When checking the prediction that $\langle M_{WD} \rangle$ among dwarf novae should be systematically higher than among the UX UMa-systems it is necessary first to remove the $\langle P \rangle$-dependence from the data. Fortunately, the dwarf nova sample (No. 6) and the sample of nova-like systems (Nos. 9 and 10) have almost the same $\langle P \rangle$ (see also Fig. 1). Therefore, deconvolution is unnecessary. Figure 1 and the numbers in Table 4 show that there is in fact a difference in $\langle M_{WD} \rangle$ of about 0.1 $M_{\odot}$ in the predicted sense.

c) The prediction that there should be no UX UMa systems below the period gap is ultimately a consequence of the disk instability model for dwarf nova outbursts (e.g. Meyer, 1985). The observational data (sample No. 11) are consistent with that model.

Table 3

Compilation of parameters of interest of 37 dwarf novae (DN) and nova-like systems (NL) that are either single- or double-lined spectroscopic binaries. m_V is the apparent visual magnitude (for DN in outburst), P the orbital period and M_{WD} the mass of the white dwarf. The additional abbreviations characterizing the type of system are UG = U Gem-star, ZC = Z Cam-star, SU = SU UMa-star (all subtypes of DN) and UX = UX UMa-star, DQ = DQ Her-star (subtypes of NL). The abbreviations used to characterize the class to which an object belongs are: E, E2 = eclipsing binary (with one or two eclipses per revolution, respectively), SB1 = single-lined spectroscopic binary, SB2 = double-lined spectroscopic binary.

References, Table 3

1 Shafter (1983)
2 Downes et al. (1986)
3 Shafter and Harkness (1986)
4 Schlegel et al. (1986)
5 Schlegel et al. (1984)
6 Dreier et al. (1986)
7 Zhang et al. (1986)
8 Wood (1986)
9 Gilliland (1982)
10 Thorstensen and Freed (1985)
11 Wood et al. (1986)
12 Wade (1986)
13 Hessman et al. (1984)
14 Hessman (1986)
15 Robinson et al. (1986)
16 Stover et al. (1981)
17 Stover (1981)
18 Horne et al. (1986)
19 Schoembs and Vogt (1981)
20 Watts et al. (1985)
21 Martin et al. (1986)
22 Young et al. (1981)
23 Penning et al. (1984)
24 Patterson (1979)
25 Chincarini and Walker (1981)

Table 3

Object	Type	Class	m_V	P(d)	$M_{WD}/M_{\odot}$	Ref.
RX And	DN,ZC	SB1	10.9	0.21154	1.14±0.33	1
V794 Aql	NL,UX	SB1	13.7	0.23	0.88±0.39	1
V1315 Aql	NL,UX	ESB1	14.4	0.139690	0.9	2
SS Aur	DN,UG	SB1	10.5	0.1828	1.08±0.40	1,3
KR Aur	NL,UX	SB1	11.3	0.16280	0.59±0.17	1
V363 Aur	NL,UX	ESB2	14.2	0.321242	0.86±0.08	4
Z Cam	DN,ZC	SB2	10.5	0.289840	0.99±0.15	1
SY Cnc	DN,ZC	SB1	11.1	0.380	0.89±0.28	1
YZ Cnc	DN,SU	SB1	10.5	0.0864	0.39±0.12	1
AC Cnc	NL,UX	ESB2	13.5	0.300478	0.82±0.13	5
OY Car	DN,SU	E2SB1	11.4	0.063121	0.90±0.04	6
HT Cas	DN,SU	E2SB1	10.8	0.073647	0.61±0.12	1,7,8
V425 Cas	NL,UX	SB1	14.5	0.1496	0.86±0.32	1
BV Cen	DN,UG	SB2	10.5	0.610116	0.83±0.10	9
WW Cet	DN,ZC	SB1	9.3	0.17578	0.50±0.14	1,10
Z Cha	DN,SU	E2SB2	11.9	0.074499	0.54±0.01	8,11,12
SS Cyg	DN,UG	SB2	8.2	0.275130	1.20±0.10	1,13,14,15
EM Cyg	DN,ZC	ESB2	12.5	0.290909	0.57±0.08	16
CM Del	DN,UG	SB1	13.4	0.162	0.48±0.15	1
U Gem	DN,UG	ESB2	9.1	0.176906	1.18±0.15	17
AH Her	DN,ZC	SB2	11.3	0.258116	0.95±0.10	18
VW Hyi	DN,SU	SB1	8.5	0.074271	0.63±0.15	19
WX Hyi	DN,SU	SB1	11.4	0.074813	0.9 ±0.3	19
T Leo	DN,UG	SB1	11.0	0.058819	0.16±0.04	1
V380 Oph	NL	SB1	14.5	0.16	0.58±0.19	1
V442 Oph	NL	SB1	12.6	0.1406	0.34±0.10	1
V2051 Oph	DN,UG	E2SB1	13.0	0.062428	0.44±0.05	20
RU Peg	DN,UG	SB2	9.0	0.3746	1.21±0.19	1
IP Peg	DN,UG	E2SB2	12.0	0.158208	0.8 ±0.2	8,21
LX Ser	NL,UX	ESB1	14.5	0.158433	0.41±0.09	1,22
SW Sex	NL,UX	ESB1	14.8	0.134938	0.58±0.20	23
RW Tri	NL,UX	ESB1	12.6	0.231883	0.44±0.08	1
SW UMa	DN,SU	SB1	9.	0.056743	0.71±0.22	1
UX UMa	NL,UX	ESB1	12.7	0.199671	0.43±0.10	1
TW Vir	DN,UG	SB1	12.1	0.182666	0.91±0.25	1
VW Vul	DN,UG	SB1	13.6	0.0731	0.24±0.06	1
AE Aqr	NL,DQ	SB2	9.8	0.411654	0.87±0.12	1,24,25

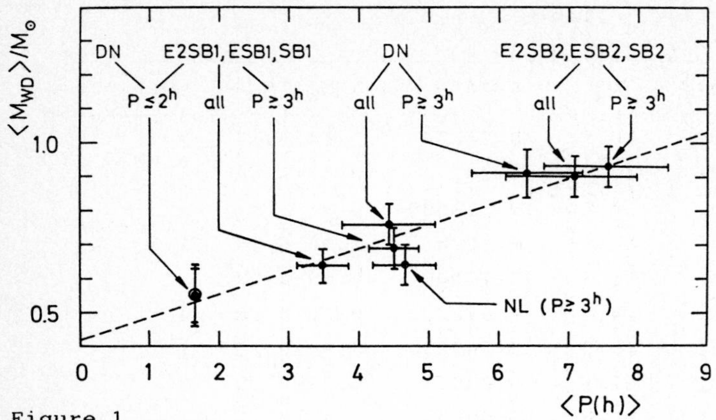

Figure 1

Dependence of the mean white dwarf mass $\langle M_{WD} \rangle$ on the mean orbital period $\langle P \rangle$ of a sample of CBs. Shown are the values of $\langle M_{WD} \rangle$ and $\langle P \rangle$ of the samples No. 1-10 of Table 4 together with the mean $\langle M_{WD} \rangle$ - $\langle P \rangle$-relation (dashed line, cf. Eq. 3). The abbreviations used to characterize the samples are explained in the caption of Table 3.

d) The prediction that $\langle M_{WD} \rangle$ should decrease with increasing m_V is one of the hallmarks of the selection effect. Ritter (1986c) has shown how this dependence arises. We test this prediction by drawing incomplete "magnitude limited" samples down to m_V = 10.0, 12.5 and 15.0, respectively (sample Nos. 12, 13, 14). Again it is important to first remove the $\langle P \rangle$ dependence. As the numbers in Table 4 show this is not necessary and we can compare the $\langle M_{WD} \rangle$ values for the different m_V directly. They do in fact show the expected trend. This trend becomes even more pronounced when the samples 12-14 are corrected for incompleteness. We have done this by giving the subsamples 18 and 19 an appropriate weight. The resulting completeness-corrected magnitude-limited samples 15-17 show now a much more pronounced decrease of $\langle M_{WD} \rangle$ with m_V. This is the strongest support, so far, in favour of our view that CBs are subject to very strong selection.

e) The last two samples (Nos. 20 and 21) in Table 4 should test whether $\langle M_{WD} \rangle$ is systematically higher in the eclipsing systems than in non-eclipsing ones. At a first glance the opposite seems to be the case. However, before a meaningful comparison can be made the fact that the two samples have a different $\langle P \rangle$ and that the eclipsing systems are systematically fainter than the non-eclipsing ones by $\Delta m_V \approx 1.5$ mag should be taken into account . When the data are corrected appropriately the difference in $\langle M_{WD} \rangle$ becomes insi-

Table 4

Defining characteristics and statistical properties of various subsamples drawn from Table 3. m_V is the limiting apparent visual magnitude, $\langle P \rangle$ and σ_P, respectively, the mean orbital period and its standard deviation, $\langle M_{WD} \rangle$ and σ_M, respectively, the mean white dwarf mass and the corresponding standard deviation. N is the number of systems in the sample. The abbreviations used in the column "sample" to characterize the sample are explained in the caption of Table 3. For further details see section 5.3.

Nr.	Sample	m_V	$\langle P(h) \rangle$	$\sigma_P(h)$	$\langle M_{WD} \rangle / M_\odot$	$\sigma_M / M_\odot$	N
1	E2SB2,ESB2,SB2		7.08±0.94	3.26	0.90±0.06	0.22	12
2	E2SB2,ESB2,SB2,$P>3^h$		7.56±0.89	2.94	0.93±0.06	0.20	11
3	E2SB1,ESB1,SB1		3.48±0.36	1.80	0.64±0.05	0.26	25
4	E2SB1,ESB1,SB1,$P>3^h$		4.50±0.36	1.43	0.69±0.06	0.26	16
5	E2SB1,ESB1,SB1,$P<2^h$		1.66±0.08	0.23	0.55±0.09	0.27	9
6	DN		4.43±0.67	3.27	0.76±0.06	0.30	24
7	DN,$P>3^h$		6.39±0.80	2.98	0.91±0.07	0.25	14
8	DN,$P<2^h$		1.67±0.07	0.22	0.55±0.08	0.25	10
9	NL		4.66±0.44	1.53	0.64±0.06	0.21	12
10	NL,$P>3^h$		4.66±0.44	1.53	0.64±0.06	0.21	12
11	NL,$P<2^h$		-	-	-	-	0
12	DN,NL	10.0	4.53±1.18	2.90	0.91±0.13	0.33	6
13	DN,NL	12.5	4.68±0.70	3.30	0.80±0.06	0.28	22
14	DN,NL	15.0	4.50±0.46	2.79	0.72±0.05	0.28	36
15	DN,NL	10.0	4.53±1.18		0.91±0.13	0.33	6
16	DN,NL (corrected)	12.5	4.72±0.70		0.77±0.07	0.26	22
17	DN,NL (corrected)	15.0	4.25±0.50		0.60±0.06	0.23	36
18	DN,NL,$10.0<m_V<12.5$		4.74±0.88	3.52	0.77±0.07	0.26	16
19	DN,NL,$12.5<m_V<15.0$		4.22±0.48	1.79	0.59±0.06	0.23	14
20	ecl. systems		4.09±0.57	2.14	0.68±0.06	0.23	14
21	non-ecl. systems		4.77±0.67	3.15	0.75±0.07	0.31	22

gnificant. Since the dependence of $\langle M_{WD} \rangle$ on the mean inclination is weak anyway, the observations are not in conflict with the predicted trend.

6. CONCLUSIONS

The main emphasis of this paper has been to examine to what extent the available observational data of white dwarfs in close binaries agree quantitatively with theoretical predictions. We have shown that if the observational data are examined carefully and, as in the case of CBs, proper allowance for the most important selection effects is made, observations and theoretical predictions agree within the uncertainties, at least as far as the mean masses of the white dwarfs is concerned. A more detailed comparison has to wait not only for many more accurate observations but also for much more detailed theoretical predictions about the frequency of occurence and the outcome of the various types of binary evolution that give birth to a white dwarf.

ACKNOWLEDGEMENTS

The author is grateful to the Max-Planck-Society for its hospitality at the Ringberg castle and to Dr. K. Butler for improving the language of the manuscript. This work was supported by the Deutsche Forschungsgemeinschaft, grant Ku 474/8-2.

REFERENCES

Chandrasekhar, S.: 1931a, Monthly Notices Roy. Astron. Soc. 91,456
Chandrasekhar, S.: 1931b, Astrophys. J. 74, 81
Chandrasekhar, S.: 1935, Monthly Notices Roy. Astron. Soc. 95, 207
Chincarini, G., Walker, M.F.: 1981, Astron. Astrophys. 104, 24
D'Antona, F.: 1982, Astron. Astrophys. 114, 289
De Grève, J.P., De Loore, C.: 1976, Astrophys. Space Sci. 43, 35
De Grève, J.P., De Loore, C.: 1977, Astrophys. Space Sci. 50, 75
Delgado, A.J., Thomas, H.-C.: 1981, Astron. Astrophys. 96, 142
Downes, R.A., Mateo, M., Szkody, P., Jenner, D.C., Margon, B.: 1986, Astrophys. J. 301, 240
Dreier, H., Barwig, H., Schoembs, R.: 1986, in preparation
Duerbeck, H.W.: 1983, Astrophys. Space Sci. 99, 363
Eddington, A.S.: 1935, Monthly Notices Roy. Astron. Soc. 95, 194
Fleming, T.A., Liebert, J., Green, R.F.: 1986, preprint
Gatewood, G.D., Gatewood, C.V.: 1978, Astrophys. J. 225, 191

Gilliland, R.L.: 1982, Astrophys. J. 263, 302

Habets, G.M.H.J.: 1986, Astron. Astrophys. 165, 95

Hamada, T., Salpeter, E.E.: 1961, Astrophys. J. 134, 683

Heintz, W.D.: 1974, Astron. J. 79, 819

Hessman, F.V.: 1986, Astrophys. J. 300, 794

Hessman, F.V., Robinson, E.L., Nather, R.E., Zhang, E.-H.: 1984, Astrophys. J. 286, 747

Holberg, J.B., Wesemael, F., Hubeny, I.: 1984, Astrophys. J. 280, 679

Horne, K., Wade, R.A., Szkody, P.: 1986, Monthly Notices Roy. Astron. Soc. 219, 791

Kippenhahn, R.: 1981, Astron. Astrophys. 102, 293

Kippenhahn, R., Weigert, A.: 1967, Z. Astrophys. 65, 251

Kippenhahn, R., Kohl, K., Weigert, A.: 1967, Z. Astrophys. 66, 58

Koester, D., Schulz, H., Weidemann, V.: 1979, Astron. Astrophys. 76, 262

Koester, D., Weidemann, V.: 1980, Astron. Astrophys. 81, 145

Koester, D., Reimers, D.: 1981, Astron. Astrophys. 99, L8

Kudritzki, R.-P., Reimers, D.: 1978, Astron. Astrophys. 70, 227

Law, W.-Y., Ritter, H.: 1983, Astron. Astrophys. 123, 33

Livio, M., Soker, N.: 1984, Monthly Notices Roy. Astron. Soc. 208, 783

Mac Donald, J.: 1984, Astrophys. J. 283, 241

Martin, J.S., Jones, D.H.P., Smith, R.C.: 1986, in: Cataclysmic Variables, IAU Coll. No. 93, J. Rahe et al. (Eds.), D. Reidel, Dordrecht, in press

Meyer, F.: 1985, in: Recent Results on Cataclysmic Variables, W. Burke (Ed.), ESA SP-236, p. 83

Meyer, F., Meyer-Hofmeister, E.: 1979, Astron. Astrophys. 78, 167

Miller, G.E., Scalo, J.M.: 1979, Astrophys. J. Suppl. 41, 513

Paczyński, B.: 1970, Acta Astron. 20, 47

Patterson, J.: 1979, Astrophys. J. 234, 978

Penning, W.R., Ferguson, D.H., Mc Graw, J.T., Liebert, J., Green, R.F.: 1984, Astrophys. J. 276, 233

Reimers, D.: 1975, in: Problems in Stellar Atmospheres and Envelopes, B. Baschek, W.H. Kegel, G. Traving (Eds.), Springer, Berlin, p. 226

Reimers, D., Koester, D.: 1982, Astron. Astrophys. 116, 341

Ritter, H.: 1976, Monthly Notices Roy. Astron. Soc. 175, 279

Ritter, H.: 1983, in: High-Energy Astrophysics and Cosmology, proc. of Academia Sinica-Max-Planck-Society Workshop on High-Energy Astrophysics, Yang Jian and Zuh Cisheng (Eds.), Science Press Beijing, Gordon and Breach Science Publ. S.A., p. 207

Ritter, H.: 1984, Astron. Astrophys. Suppl. Ser. 57, 385

Ritter, H.: 1986a, in: The Evolution of Galactic X-Ray Binaries, J. Trümper, W.H.G. Lewin, W. Brinkmann (Eds.), D. Reidel, Dordrecht, p. 207

Ritter, H.: 1986b, Astron. Astrophys., in press

Ritter, H.: 1986c, Astron. Astrophys., in press

Ritter, H.: 1987, Catalogue of Cataclysmic Binaries, Low-Mass X-Ray Binaries and Related Objects, 4th. edition, in preparation

Ritter, H., Burkert, A.: 1985, in: Recent Results on Cataclysmic Variables, W. Burke (Ed.), ESA SP-236, p. 17

Ritter, H., Burkert, A.: 1986, Astron. Astrophys. 158, 161

Ritter, H., Özkan, M.T.: 1986, Astron. Astrophys., in press

Robinson, E.L., Zhang, E.-H., Stover, R.J.: 1986, Astrophys. J. 305, 732

Salpeter, E.E.: 1955, Astrophys. J. 121, 161

Schatzmann, E.: 1958, White Dwarfs, North Holland, Amsterdam

Schlegel, E.M., Kaitchuck, R.H., Honeycutt, R.K.: 1984, Astrophys. J. 280, 235

Schlegel, E.M., Honeycutt, R.K., Kaitchuck, R.H.: 1986, Astrophys. J. 307, 760

Schoembs, R., Vogt, N.: 1981, Astron. Astrophys. 91, 25

Schoenberner, D.: 1979, Astron. Astrophys. 79, 108

Schoenberner, D.: 1981, Astron. Astrophys. 103, 119

Schoenberner, D.: 1983, Astrophys. J. 272, 708

Shafter, A.W.: 1983, Ph.D. Thesis, Univ. California, Los Angeles

Shafter, A.W., Harkness, R.P.: 1986, Astron. J. 92, 658

Stover, R.: 1981, Astrophys. J. 248, 684

Stover, R.J., Robinson, E.L., Nather, R.E.: 1981, Astrophys. J. 248, 696

Strand, K.A.: 1951, Astrophys. J. 113, 1

Strand, K.A.: 1977, Astron. J. 82, 745

Thorstensen, J.R., Freed, I.W.: 1985, Astron. J. 90, 2082

Trimble, V.: 1983, Nature 303, 137

Wade, R.A.: 1986, in: Cataclysmic Variables, IAU Coll. No. 93, J. Rahe et al. (Eds.), D. Reidel, Dordrecht, in press

Wali, K.C.: 1982, Physics Today, October 1982, p. 33

Warner, B.: 1973, Monthly Notices Roy. Astron. Soc. 162, 189

Warner, B.: 1976, in: Structure and Evolution of Close Binary Systems, IAU Symp. No. 73, P. Eggleton, S. Mitton, J. Whelan (Eds.), D. Reidel, Dordrecht, p. 85

Watts, D.J., Bailey, J., Hill, P.W., Greenhill, J.G., Mc Cowage, C., Carty, T.: 1985, Astron. Astrophys. 154, 197

Webbink, R.F.: 1979, in: White Dwarfs and Variable Degenerate Stars, IAU Coll. No. 53, H.M. van Horn, V. Weidemann (Eds.), Rochester, University of Rochester, p. 426

Webbink, R.F.: 1985, in: Interacting Binary Stars, J.E. Pringle, R.A. Wade (Eds.), Cambridge University Press, p. 39

Webbink, R.F.: 1986, in: Critical Observations vs. Physical Models for Close Binary Systems, K.C. Leung, D.S. Zhai (Eds.), Gordon and Breach, New York, in press

Wegner, G.: 1980, Astron. J. 85, 1255

Weidemann, V.: 1984a, Astron. Astrophys. 134, L1

Weidemann, V.: 1984b, paper presented at the 5th European Workshop on White Dwarfs, Kiel 1984, unpublished

Weidemann, V., Koester, D.: 1983, Astron. Astrophys. 121, 77

Weidemann, V., Koester, D.: 1984, Astron. Astrophys. 132, 195

Wood, J.: 1986, in: Cataclysmic Variables, IAU Coll. No. 93, J. Rahe et al. (Eds.), D. Reidel, Dordrecht, in press

Wood, J., Horne, K., Berriman, G., Wade, R.A., O'Donoghue, D., Warner, B.: 1986, Monthly Notices Roy. Astron. Soc. 219, 629

Young, P., Schneider, D.P., Shectman, S.A.: 1981, Astrophys. J. 244, 259

Zhang, E.H., Robinson, E.L., Nather, R.E.: 1986, Astrophys. J. 305, 740

Part IV

Active Galactic Nuclei

Active Galactic Nuclei: Ten Theoretical Problems

M.J. Rees

Institute of Astronomy, Madingley Road, Cambridge CB3OHA, UK

The physics of active galactic nuclei (AGNs) and related objects involves hypersonic (and perhaps relativistic) gas dynamic, in an environment where radiation fields may be very intense. Various aspects of such phenomena have been reviewed elsewhere (e.g.[1-3]) and I shall not repeat previous discussions in the present written summary of my talk. An occasion like this is perhaps an opportunity to look forward into the future, as well as to recall what has already been accomplished. I shall therefore address some topics that seem ripe for further attention from a theoretical point of view, focussing especially on problems where the computational expertise and facilities of the Munich group could be a special asset. The ten problems I have chosen are a rather arbitrary selection; and, needless to say, there are many further questions that must be settled before our understanding of AGNs is as firmly based as even our present theories of stellar structure and evolution.

1. The structure of the broad line-emitting region

The best-studied aspect of AGNs is the region that emits the broad lines. Here, the level of ionization is a direct measure of the ratio of radiation pressure and gas pressure, and implies that radiation pressure can be dynamically important. A generic feature of AGNs and related objects is the dominance of radiation pressure, and its possible dynamical effects in driving an outflow of material. Among the well-known consequences of this, discussed especially by Professor Kippenhahn and his collaborators [4-6], is the possibility that radiation pressure can accelerate photoionized clouds away from the central source of UV continuum.

But the dynamics of these clouds are still perplexing -- we do not yet know how they are moving, let alone what mechanism drives them (for a recent survey see MATHEWS [7]. Indeed it is unclear whether there are discrete clouds surrounding a central ionizing continuum, rather than the line-emitting gas being concentrated in a disc irradiated from outside its plane [8]. Studies of time-variation in line profiles are now starting to discriminate among inflow, outflow and 'orbiting cloud' models; but we are still unclear about the cloud sizes and shapes, and how long individual clouds persist (there is no necessity for a cloud to last even for one orbital time). Moreover, the origin of the emitting gas, its spatial configuration (disc or not?), how (if at all) it is confined and the degree of isotropy of the continuum radiation, are equally uncertain.

2. The properties of gas clouds closer to the central object

The radiation energy density in the broad-line region is well below that of a black body at 10^4K. The clouds, with electron temperature $\sim 10^4$ K, are therefore exposed to dilute radiation -- were this not so, approximate LTE would prevail, and the emission lines would not stand out above the continuum intensity. But there may be 3 or 4 powers of ten difference in scale between the emission line region and the central object, so it is interesting to inquire whether clouds could exist (with similar covering factor) at a range of smaller radii r around a central continuum source. If the density varied as r^{-2}, the ionization parameter would be independent of r; however at higher densities the usual HII region assumptions progressively break down. When the densities are sufficiently high, each cloud behaves like a black body (or, more precisely, like a segment of an irradiated stellar atmosphere). The often-observed UV bump, indicating a thermal black body component, is conventionally attributed to an accretion disc. However, all we really know is that there is a surface area radiating thermally -- it could be a lot of small clouds rather than single surface.

It is not obvious that one can exclude the possibility that clouds exist (with a similar covering factor) at all logarithmic intervals of radius. Theorists who have offered reasons why emission line clouds may form at a special radius $r \cong 1$ pc (e.g. because a wind from the central object is stopped by a shock at this distance, or because this is where the form of the potential changes from 1/r to a shallower r-dependence determined by the stellar distribution) may be trying harder than they need.

The surface radiation from optically thick clouds and the inner parts of accretion flows (and of course from supermassive stars) would resemble that from ordinary massive O or B stars. There are just two possible differences:

(i) The dominance of electron scattering opacity is even greater, with the result that there may be an even greater disparity between bolometric and colour temperatures.

(ii) The surface gravity is not enormously larger than for ordinary stars, but the specific gravitational binding energy is very large, and so therefore is the escape velocity V_{esc}. This means that, if fields were to grow via differential rotation to a strength anywhere near equipartition, the Alven speed ($\sim V_{esc}$) would be high enough for magnetic fields to induce much stronger flare-like activity than in ordinary stars.

Magnetic fields are inevitable in accretion flows, and one of the hardest things to predict is the relative importance of thermal and non-thermal radiation mechanisms. The only trends that one can predict with any confidence are that non-thermal effects would tend to be more significant than in stars (because there is a deeper gravitational potential well to confine strong fields), and would be of greater relative importance for low $\dot{M}$ (because cooling by thermal electrons is inefficient at low densities). Three modes of radiation need to be considered in the central regions of AGNs; depending essentially on the ratio of the energy dissipation rate per unit volume to the plasma density:

(a) When the density is low (or the dissipation rate very high), the only radiative processes with adequate efficiency will be those involving relativistic electrons.

(b) In intermediate cases, subrelativistic electrons may be able to radiate with adquate efficiency via bremsstrahlung and Comptonisation.

(c) For high densities, the radiation approaches a black body; this happens only when the optical depth due to Thomson scattering is $\gtrsim 10^3$.

Regime (c) is not attained in accretion flows unless $\dot{M}$ is large, and/or the effective viscosity is low enough to permit a long residence time for material before it gets swallowed by the central object. In general, we might expect a three-phase structure around massive black holes: dense optically-thick clouds or filaments may be embedded in a hotter (but non-relativistic medium); relativistic particles accelerated by shocks may constitute a third phase, contributing importantly to the radiative output even though the fraction of particles in this phase could be small. The possibility of a pair-dominated plasma [9] is a further complication.

3. Transport processes when radiation pressure is overwhelmingly dominant: convection and 'photon bubbles'

Internal conditions in accretion flows may approach thermal equilibrium (case (c) above), especially in thick discs or "donuts" where $\dot{M}$ is high and the effective viscosity low [10] (see section 7). In the opposite limit of low densities and rapid inflow, the cooling may be so inefficient that gas can remain at the virial temperature [11]. The ratio of radiation and gas pressure at a large optical depth depends on the value of T_{eff}^3/n., where T_{eff} is the effective temperature and n the particle density. Inside stars, the mass-dependence of this ratio is well-known. For typical stars, where radiation pressure support is unimportant, it goes as M^2; but for very massive stars ($M \gtrsim 100\ M_\odot$) radiation pressure is dominant, and the ratio then rises roughly as $M^{1/2}$. This means that for supermassive stars such as might be a stage in AGN evolution, the radiation pressure contributes more than 99 per cent of the total.

The dominance of the radiation pressure is even more extreme in accretion flows. The difference between these and supermassive stars is that the density n can be very much lower, the binding mass being not the gas itself (whose active gravitational effects are generally negligible) but a central collapsed body. The ratio of radiation and gas pressure can now be estimated by the following simple argument. There are two characteristic temperatures: one is

$$T_{virial} \simeq 2\times10^{12} \ (r/r_{Schw})^{-1} \ K \tag{1}$$

the other is the effective (black body) temperature

$$T_{eff} \simeq 10^{6}(M_h/10^{8} \ M_{\odot})^{-1/4}(r/r_{Sch})^{-1/2}[\tau(>r)]^{1/4} \ K \tag{2}$$

In this approximate expression (for a quantity which is in general geometry-dependent) τ is a measure of the optical depth outside a radius r. If material with temperature T_{eff} is to be supported in a potential well the energy per proton must be of order kT_{virial}; this therefore requires that radiation pressure (or photon density) must exceed gas pressure (or proton density) by (T_{virial}/T_{eff}). Ratios of order 10^6 are now possible. (Such dominance occurs, incidentally, in another very different context: the early universe, where in the fireball before recombination, radiation pressure dominates gas pressure by the photon/baryon ratio, known to be $\gtrsim 10^8$.)

In the optically thick flows associated with supercritical accretion, photon diffusion yields a luminosity L_{Edd}; but it is interesting to ask whether convective-type processes can transport energy much more efficiently and thereby blow off a strong wind or jet-like outflow above the surface (cf. section 8). Little attention seems to have been given to the nature of convection in these contexts. But one can readily see that, insofar as radiation pressure dominates, conventional convective transport becomes <u>inefficient</u>: this is because, when the pressure is predominantly due to radiation (and proportional to T_{eff}^4) <u>pressure balance</u> implies close <u>temperature equality</u>. Specifically, if we compare the temperature of rising (underdense) and falling (overdense) elements, then the temperature difference is is

$$\left|\frac{\delta T}{T}\right| \simeq (T_{eff}/T_{virial}) \left|\frac{\delta \rho}{\rho}\right| \tag{3}$$

We must also ask how rapidly inhomogeneities would be erased. A minimum mixing rate is set by thermal diffusion. The main transport process arises from the photons: regions of high baryon density are slightly cooler; so photons tend to diffuse into cooler regions and inflate them until the baryon density is homogenised. Because the temperature differences are small (equation (3)), the time taken for thermal diffusion to erase density inhomogeneities is, however, (T_{virial}/T_{eff}) longer than the timescale for a photon to random walk across the relevant scale. Of course, mixing could occur on larger scales via macroscopic proesses - for instance, Kelvin-Helmholtz instabilities at the boundaries between rising and falling elements.

To transport energy at a given rate via convection, the entropy gradient must be $\sim(T_{virial}/T_{eff})$ times greater than it would need to be if the pressure were all contributed by gas.

The radiation can be envisaged as a light fluid intermixed with a heavier fluid (the matter), and this suggested to PRENDERGAST and SPIEGEL [12] an analogy with fluidised beds. However it is not clear how valid this is: in a fluidised bed, both substances independently display fluidlike behaviour; but a photon gas is not in itself fluidlike - it resembles a fluid only because of repeated scatterings by the electrons.

Insofar as we can neglect small-scale diffusion, bubbles with $|(\delta\rho/\rho)| \simeq 1$ would rise on a timescale

$$\left|\frac{\delta\rho}{\rho}\right|^{-1} \times \left\{\frac{\text{bubble size}}{\text{scale height}}\right\}^{-1/2} t_{dyn}$$

As a corollary, the regions of above-average baryon/photon ratio would sink at a similar rate. Thus it would be possible for the bulk of the radiation energy to escape in one dynamical timescale if the density fluctuations had an amplitude of order unity. This corresponds to a rate of energy release higher by a factor (T_{virial}/T_{eff}) than could ever result from conventional convection. The essential difference between the two cases is that the 'bubbles' are assumed to break through the surface without losing their identity; convective eddies, on the other hand, lose their identity in one turnover time.

4. Radiation pressure in jets: line-locking, etc.

There is clear evidence that, on the extragalactic scale, collimated jet-like outflows transport energy from AGNs out to extended radio sources (see Biermann's contribution to these proceedings). Understanding the nature and origin of these jets is essential to an understanding of active galaxies. The mildly relativistic jets in SS433 may be a miniature version of some jet-like phenomena in extragalactic radio sources. In some respects, they are more easily studied and more tightly constrained than those in any extragalactic object. Only in SS433, for example, do we directly measure the speed of outflow, by observing the time-varying Doppler shifts of the emission lines. These measurements indicate that this has remained remarkably constant at .26c, as well as being uniform across the jet. This steady speed places a powerful constraint on all models.

A mechanism that naturally explains both the value and the constancy of the jet outflow speed is 'line-locking'; this was first suggested in the context of

SS433 by MILGROM [13]. It requires the following: (1) an underlying continuum flux strong enough for radiation pressure to overcome gravity; (2) that the dominant momentum transfer is through Lyman-line absorption by some hydrogenic ion; and (3) that the continuum flux falls off sharply above the Lyman edge for that ion (as occurs, for example, in early-type stars at the H Lyman edge). Gas would then be accelerated up to a terminal velocity such that the relativistic Doppler shift of the accelerated gas with respect to the underlying continuum sources shifted the Lyman-edge wavelength down to the local comoving Lyman α wavelength. This happens for a velocity of 0.28c (actually at a somewhat lower velocity because blanketing by higher Lyman lines would attenuate the continuum at slightly longer wavelengths than the Lyman limit).

The general suggestion of line-locking is not new. There is convincing evidence of 'locking' between closely-spaced lines in some stellar spectra. Studies of radiation-driven acceleration in the quasar context were stimulated by various claims to have found preferred ratios between different redshift systems in quasars displaying multiple absorption lines. The reality of this effect is still controversial, though there is some recent evidence for it.[14]

In the 1970s Professor Kippenhahn and his colleagues studied radiation-driven winds from quasars, assuming a power-law photoionization continuum and taking the ionization equilibrium consistently into account [4-6]. They concluded that radiative forces can plausibly generate speeds comparable with those of the broad-line clouds, but that relativistic velocities could not readily be attained. (In any case, it now seems clear that most absorption lines in quasar spectra whose redshifts differ substantially from the emission redshift arise from intervening intergalactic material physically unrelated to the quasar).

Radiative acceleration via the Lyman lines in SS433 has been more fully explored by SHAPIRO et al. [15]. An important constraint is that the acceleration cannot start too close to a central compact object because the intensity (and, correspondingly, T_{eff}) would then be too high to permit hydrogen to survive in neutral form. This problem is evaded if the acceleration is attributed to a high-z hydrogen like ion (Fe, for instance); but there is then a loss of efficiency because the abundances of such ions (and the appropriate cross sections) are lower. Although the line-locking mechanism cannot be ruled out, it is fair to say that fuller investigation has made it seem less likely; it is certainly hard to incorporate the necessary constraints in any astrophysically-plausible model for this unusual system.

SS433 offers the clearest evidence of thermal matter being expelled from an object whose photon luminosity probably exceeds L_{Edd} - even if the ejection speed did not have the special value suggestive of line-locking, it would still be the most likely object, of all those known, where radiative driving might be important. BEGELMAN and I [16] have considered whether the jets could be accelerated by thermal continuum radiation, Compton scattering then being the relevant opacity.

For this to work, some of the space around a compact object must be filled with a mixture of matter and radiation such that $(p/\rho)^{1/2}$ exceeds the escape velocity; if the surrounding material were in a rotationally-flattened distribution, the buoyant material would be expelled in 'twin exhaust' jets along the minor axis rather than in the uncoordinated photon bubbles described in section 3. A volume with high (p/ρ) would naturally arise within the magnetopause of a neutron star: the pressure holding up surrounding material can here be supplied by magnetic stresses [16].

In this latter mechanism, the acceleration would occur close to the central object, where radiation would be trapped - multiple scattering then permits efficient conversion of radiant energy into bulk kinetic energy (in contrast to the optically thin case, when each photon can be used only once, and its momentum, not energy, is the relevant quantity). The details would be influenced by radiative viscosity, which would tend to entrain matter into the jet. More detailed computations are needed in order to calculate the velocities of these jets, and how sensitive they are to ambient conditions (these points are both very important in the application to SS433).

Radiation pressure acceleration is unlikely to be relevant to the jets in strong radio galaxies. In the nuclei of these galaxies, the photon luminosity is low, and the main theoretical problem is to explain how the energy from the AGN can be channelled predominantly into the low-entropy form of a relativistic outflow and/or Poynting flux; electromagnetic processes around a spinning hole (with low accretion rate $\dot{M}$) offer, I believe, the most promising possibility [11]. However it is worth noting that if there were SS433-like jets in radio-quiet quasars, they might have escaped our attention; the time-scale for precession would be too slow for us yet to have detected it, and the only signature would then be weak emission lines with a redshift discordant from the quasar itself.

5. Non-thermal jets on scales <1 pc

The superluminal components observed at milli-arc-second resolution by VLBI have (deprojected) lengths of $\sim 10^{20}$ cm - 10^{-4} of typical VLA map scales - and imply relativistic bulk motion on these scales. There are, however, persuasive reasons for attributing the primary energy production to relativistically deep potential wells on scales 10^{14} - 10^{15} cm. Moreover, even though we have no direct evidence that well-collimated jets exist on scales below those accessible to VLBI, the initial bifurcation and collimation must be imposed on these small scales if the long-term stability of the jet axis in extended sources is due to the gyroscopic effect of a spinning black hole (cf. section 6). It is important not to forget that many powers of ten difference in scale are involved: if collimation is initiated on scales of 10^{15} cms or less, the jets may face many vicissitudes before they penetrate to the much larger distances where we observe the radio phenomena. They may be destroyed and recollimated; they could even change direction. The small scale jet may be lined up with the direction of the rotation axis of the central massive object while the large scale one may lie along the rotation axis of the galaxy.

The stuff ejected from 10^{15} cm may be electron-positron plasma rather than "ordinary" electron-ion plasma. Although there have been several recent investigations of cooling and radiative transfer in such plasmas (reviewed by SVENSSON [9], little attention has yet been given to their dynamics. Energy can also be transported via Poynting flux - either as a large-scale field carried out with the particles (a directed MHD wind) or as low-frequency wave modes - and this can, in principle, swamp the kinetic energy carried by the charged particles themselves.[17,18]

The flow patterns on the unobservably small scales 10^{15} - 10^{19} cm, if we could probe them in the same detail that the VLA provides for scales a million times larger, would no doubt prove just as complex: there would be a multiphase medium (see section 2), entrainment of surrounding gas, bending by transverse pressure gradients, and shocks where the jet impinges on the dense gas clouds that emit the broad emission lines. But one general statement can be made. The flow patterns would not simply be a scaled down version of those seen on larger scales, because one key number - the ratio of radiative cooling times ($\propto r^2$ for a simple diverging jet) to dynamical times ($\propto r$) - is proportional to r rather than being scale-independent. Consequently, the flows on small scales would tend to be less elastic and more dissipative; they are less likely to maintain a high internal pressure, and would dissipate more energy if bent through large angles.

Progress in this area will surely depend on increasingly sophisticated hydrodynamical codes. Two-dimensional codes have already uncovered some gas dynamical properties of supersonic flows that were unanticipated by analytical models and may have counterparts in radio maps. We must await 3D codes before we can hope to simulate the non-linear development of instabilities and the bends in jets. (Some limited insight may be gleaned from the combined use of 2-D Cartesian and cylindrical simulations). The other important computational development will be the use of MHD codes. Only in this way will we be able to see if it really is practical to confine jets magnetically, whether or not the polarization patterns observed in jets can be explained in terms of the kinematics of expanding shear flows (and to study flow patterns around black holes, as discussed further in the next sections).

6. Lense-Thirring precession in thick accretion flows

Many features of AGNs merely indicate a deep gravitational potential well, but do not depend on whether the field exactly resembles what general relativity predicts for the metric around black holes. Special interest attaches, therefore, to ways whereby AGN observations could "diagnose" the nature of the metric, and thereby offer tests of general relativity beyond the weak-field domain to which all current tests are restricted. The most dramatic possibility is that the power supply in radio galaxies (whose output is primarily in the form of relativistic plasma beams) could come directly from the spin-energy of a Kerr black hole, rather than directly from accretion [11,18]. But there is another intrinsically relativistic effect that can perhaps only be studied by 3-D computations: this is Lense-Thirring precession. Near to a spinning (Kerr metric) black hole, this precession would tend to make the flow axisymmetric relative to the hole, even if the angular momentum of infalling matter (debris form tidally-disrupted stars, etc?) were misaligned with the hole's. A simple often-used argument, first given by BARDEEN and PETTERSON [19] is that the spinning hole enforces axisymmetry out to the radius r_{BP} where the precession time $t_{L-T} = t_{inflow}$. If valid, this argument suggests that in optically thick discs, which must have a low "α", the radius r_{BP} is large. Several authors (inlcuding the present one) have presumed that this effect renders the assumption of axisymmetry in the relativistic domain close to the hole physically realistic rather than just a mathematical simplification, provided that the viscosity is reasonably small. Such a simplification is of course welcome - and, indeed, almost a prerequisite for any analytical progress in studying (for instance) electromagnetic extraction of energy from spinning holes.

However, PAPALOIZOU, PRINGLE and KUMAR [20,21] point out that Bardeen and Petterson overlooked an extra contribution to the angular momentum transport in a twisted disc. In slightly-tilted thin discs this extra torque, a consequence of the small azimuthal component of the pressure gradient, makes r_{BP} decrease as "α" decreases. This pressure-induced effect may not invalidate the Bardeen-Petterson criterion for thin discs when the angle of tilt is large; but could very well be important in thick discs or "donuts". Elucidation of this problem may have to await 3D computations, and perhaps also a better understanding of what actually causes the viscosity. If the Lense-Thirring precession were effective, the resultant gyroscopic' effect near the hole would guarantee a constant orientation for jets collimated close to the hole irrespective of the provenance of the infalling matter; but were the axisymmetric approximation actually to be invalid, long-term stability could not be achieved unless the infalling material were supplied with an invariant angular momentum vector, or unless the "nozzles" that collimate extended jets had a much larger scale.

7. Instabilities in accretion flows

Even if material accretes steadily, with constant angular momentum, the resultant primary radiation output is sensitive to uncertain details of the flow. The well-known "donut" configurations, analysed primarily by Abramowicz and collaborators, [10] require a high value of M are supported "vertically" by radiation pressure, and radiate $L \simeq L_{Edd}$. If the effective viscosity were very low (and the storage time in the donut correspondingly high) the material would be dense enough to thermalise the outgoing radiation. However, the thermalisation requirement is a very stringent one. If we define "α" = $(v_{inflow}/v_{free\ fall})$, then one requires "$\alpha$" $\lesssim 10^{-3}$ $(\dot{M}/\dot{M}_{crit})$ to thermalise radiation even at the densest part of the donut ($\dot{M}_{crit}$ being defined as L_{Edd}/c^2). A still lower value of "α" would be needed in order to maintain thermalisation out to a photosphere at much larger radius (because the density must, for a stable entropy gradient, fall off at least as steeply as r^{-3}).

We do not know what "α" is likely to be. However, these structures are vulnerable to non-axisymmetric instabilities [22], which may expel angular momentum so efficiently that they maintain an effective "α" far too high to permit thermalisation. So the inflowing material may behave in an irregular "cauldron-like" fashion: a long time exposure would still reveal a somewhat flattened axisymmetric configuration (though not necessarily with a completely empty "funnel" along the spin axis), but large irregular non-axial motions would make the density distribution and flow pattern variable on a dynamical timescale. The radiation from such a "cauldron" could be predominantly Comptonised bremsstrahlung or non-thermal radiation (see section 2); it would not, however, approximate a black body spectrum.

It is important to find out how these instabilities depend on the angular momentum distribution. Such information would reduce the arbitrariness of current models for "donuts" in two interconnected ways. First, it may contrain the angular momentum distribution $\alpha(r)$ (which is at the moment a free parameter within the limits $\alpha \propto r^2$ and $\alpha \simeq$ constant); second, these instabilities may provide the dominant effective viscosity, and thereby set a lower limit on "α".

8. Unsteady accretion

Realistically, one expects accretion flows to display instabilities on all timescales - from the minimum dynamical time r_{Schw}/c, right up to the overall duration of the AGN phenomenon. Of particular interest are cases when the material is supplied unsteadily. For instance:

(i) The gas may be sporadically supplied as debris from tidally-disrupted stars, which forms an axisymmetric 'donut', and then evolves under the action of viscosity.

(ii) If a massive black hole were to form via post-Newtonian instability of a supermassive star which was slowly rotating, then some fraction of the original material might have too much angular momentum to fall directly into the hole.

Either of the above processes could give rise to a 'donut' surrounding the hole (which could, in case (ii) have a mass comparable to the hole itself).

The subsequent evolution of such systems depends on the ratio of two timescales: the first of these is the timescale t_{visc} for viscous redistribution of angular momentum; the second is the time t_{rad} taken to radiate the gravitational binding energy at luminosity L_{Edd}. If $t_{visc} > t_{rad}$ the original donut would deflate to a thin disc or ring (which might then become selfgravitating). In the opposite case, where $t_{visc} < t_{rad}$ the structure would remain physically thick: during a viscous timescale the resultant donut would evolve through a family of configurations with almost equal overall binding energies but different angular momentum distributions: some material would fall into the hole, while the rest could be blown off as a supercritical wind or jets.

9. Evolution of precursors of massive black holes

One can readily envisage various evolutionary pathways whereby a galactic nucleus may run away towards gravitational collapse. It is unclear which route is the more common one: nor do we know how (nor, indeed, whether) specific states or routes can be associated with particular types of violent activity in galactic nuclei. Two things only seem clear:

(i) Most evolutionary pathways involve an inexorably deepening potential well, and it is hard to avoid concluding that much of the material ends up by collapsing into a black hole.

(ii) Some forms of activity are better explained by processes involving accretion onto an already-formed black hole, or energy extraction from a spinning hole, than in terms of any "precursor" stage.

The various "precursor" stages are nevertheless worth much more study; moreover, they pose many problems which require large-scale computation.

(i) Evolution of dense star clusters. When the velocity dispersion in a cluster of ordinary stars approaches ~1000 km s^{-1}, stellar collisions become important. What happens next is still controversial, despite the many papers on the topic, basically because the relative importance of stellar disruption and stellar coalescence is uncertain. Realistic models of stellar collisions would certainly be a great help in elucidating this, and in deciding whether a dense star cluster would give rise to ~100 $M_{\odot}$ coalesced stars, leading to multiple supernovae (leaving behind a cluster of neutron stars or black holes?); or whether the stars are all destroyed, leaving an amorphous gas cloud which evolves into a single supermassive star. A cluster of neutron stars could develop a dense core that eventually collapses to a black hole; and a supermassive star would be vulnerable to post-Newtonian dynamical instability.

(ii) Dynamical stability and evolution of supermassive stars. It could be interesting to compute models of supermassive stars with realistic differential rotation. If such an object goes dynamically unstable, its core collapses to a black hole, but a substantial fraction of the original mass may find itself with too much angular momentum to be swallowed. This material would form a "donut", supported by radiation pressure, which would evolve on a Kelvin or viscous timescale (whichever is shorter). These (possibly self-gravitating) tori, would undergo secular evolution rather than being stationary (cf. section 8). They may nevertheless be sufficiently common, long-lived and efficient to provide models for some category of radio-quiet quasars.

It is often argued that radiation pressure restricts the rate at which black holes can grow by accretion, enforcing a growth timescale $\gtrsim 10^8$ yrs. This argument is, however, unconvincing. There is no necessity that any specific amount of energy be radiated per unit mass swallowed: "donuts' permit steady accretion with arbitrarily high $\dot{M}$ and low efficiency; spherical accretion with high $\dot{M}$ leads to a large "trapping radius", so that most of the radiation is advected into the hole rather than escaping; and an entire super-massive star which goes dynamically unstable can collapse to a black hole on a free-fall timescale, without needing to radiate any energy at all. We shall need to understand these processes better before we can make more reliable estimates of the expected masses of black holes (defunct remnants of past AGNs) in typical present-day galaxies.

10. Stellar disruptions by black holes

Another set of phenomena arise when stars pass close enough to a massive black hole to be tidally distorted and even disrupted. (for a given type of star, the tidal radius around a hole of mass M_h scales as $M_h^{1/3}$; consequently, solar-type stars cannot be disrupted by holes exceeding 10^8 $M_\odot$ without being swallowed first.)

Tidal disruption of stars is a widely discussed fuelling mechanism for black holes - a star on a highly eccentric orbit is disrupted, and the debris then evolves on a timescale controlled by viscous effects. The observational consequences depend on the answers to three interlinked questions:

(a) What fraction of the debris goes down the hole, rather than being expelled?

(b) What is the radiative efficiency for the accretion process? In other words, how many ergs of energy are radiated for each gram that is swallowed?

(c) How long does it take to "digest" one star? In particular, how does the "flare duration" and decay timescale for such a process compare with the interval between one stellar disruption and the next?

The expulsion can happen in several ways. The debris may form a donut within which viscosity liberates energy at a supercritical rate (see section 8); a small fraction of the mass, accreted with high efficiency, could then yield enough energy to eject the remainder. Alternatively, material can be expelled before an axisymmetric configuration has a chance to form. When a star passes within the tidal radius and is disrupted, its self-binding energy must be supplied by the orbital motion, the debris is, on average, more bound to the hole than was the original stellar orbit. But shocks occurring during the disruption could result in a "spray" of debris on a wide range of orbits (with a consequent spread of specific binding energies to the hole), so that some fraction escapes on hyperbolic orbits; and if the sudden tidal compression triggered explosive nuclear reactions, the energy released during the 'flyby' might exceed the star's self-binding energy, so that almost all the debris could escape from the hole's gravitational influence.[23]

Processes such as these are important in all galactic nuclei, and may have readily observable consequences in our Galactic Centre if a 10^6 $M_\odot$ hole lurks there - indeed, they could be responsible for the peculiar arm-like features in the central 2 pc of our galaxy.[24] They involve many of the same complexities as calculations of stellar evolution and supernovae, but are also intrinsically non-axisymmetric, and therefore may require a fully 3-dimensional treatment.

I am most grateful to the organisers of this meeting for extending me an invitation, and an opportunity to offer personal greetings to Professors Kippenhahn and Schmidt on this anniversary occasion.

REFERENCES

1. M.C. Begelman, R.D. Blandford and M.J. Rees, Rev. Mod. Phys. 56, 255 (1984)
2. M.J. Rees, Ann. Rev. Astr. Astrophys. 22, 471 (1984)
3. P. Wiita, Phys. Reports 123, 117 (1985)
4. R. Kippenhahn, J.J. Perry and H-J. Röser, Astr. Astrophys. 34, 211 (1974)
5. R. Kippenhahn, L. Mestel and J.J. Perry, Astr. Astrophys. 34, 123 (1975)
6. R. Kippenhahn, Astr. Astrophys. 55, 125 (1979)
7. W.G. Mathews, in "Radiation Hydrodynamics in Stars and Compact Objects", eds. D. Mihalas and K-H. Winkler (Springer-Verlag, Berlin) p.346 (1986)
8. H. Netzer, preprint
9. R. Svensson in "Radiation Hydrodynamics in Stars and Compact Objects", eds. D. Mihalas and K-H. Winkler (Springer-Verlag, Berlin) p.325 (1986)
10. M. Abramowicz, M. Jaroszynski and M. Sikora, Astrophys. J. 242, 772 (1980)
11. M.J. Rees, M.C. Begelman, R.D. Blandford and E.S. Phinney, Nature 295, 17 (1982)
12. K.H. Prendergast and E.A. Spiegel, Comm. Astrophys. Sp. Phys. 5, 43 (1973)
13. M. Milgrom, Astr. Astrophys. 78, L.9 (1979)
14. C. Foltz, F. Chaffee, S. Morris and R. Weymann, Steward Observatory Preprint (1986)
15. P. Shapiro, M. Milgrom and M.J. Rees, Astrophys. J. Suppl. 60, 393 (1986)

16. M.J. Rees in "VLBI and Compact Sources" eds. R. Fanti and G. Setti (Reidel) (1984)

17. M.C. Begelman and M.J. Rees, M.N.R.A.S. 206, 209 (1984)

18. E.S. Phinney, Unpublished Cambridge Ph.D. Thesis (1983)

19. J.M. Bardeen and J.A. Petterson, Astrophys. J. (Lett) 195, L.65 (1975)

20. J. Papaloizou and J.E. Pringle, MNRAS 213, 799 (1985)

21. S. Kumar and J.E. Pringle, MNRAS 213, 435 (1985)

22. J. Papaloizou and J.E. Pringle, MNRAS 208, 721 (1984)

23. B. Carter and J.P. Luminet, Nature 296, 211 (1982)

24. M.J. Rees in "The Galactic Center" ed. D. Backer and R. Gentzel (AIP, New York) (1987).

Observations of a Complete Sample of Relativistic Sources in Active Galactic Nuclei *

P.L. Biermann

Max-Planck-Institut für Radioastronomie, Auf dem Hügel 69, D-5300 Bonn 1, Fed. Rep. of Germany

We discuss two independent complete samples of radio sources for which extensive VLBI observations exist. The first sample is based on the Jodrell Bank 966 MHz survey and comprises 30 radio lobe dominated sources. Since radio lobe emission is accepted to be isotropic, this sample should be isotropically distributed with respect to the observer. And yet there seem to be more sources with apparent superluminal motion in their core than compatible with an isotropic distribution. The second sample is based on the Effelsberg 5 GHz survey north of declination 70° (S5) and comprises 13 flat spectrum radio sources. Here the evidence from VLBI, radio variability, optical polarization and X-ray observations suggests that all sources in the sample show the effects of bulk relativistic motion in their nuclei. Combining the results from both samples leads to the suggestion that extended radio sources with strong heads like 3C179, 3C390.3 and S5 1928+73 maintain bulk relativistic motion all the way from the core to the strong emission knot in the head where multiple oblique shocks may dissipate the energy (Norman et al. 1984, Lind and Blandford 1985). The success of finding apparent superluminal motion in the 966 MHz sample is thus due to a selection effect which biases the sub-sample actually observed due to relativistic boosting mostly in the cores and to a small degree in the heads.

INTRODUCTION

Flux density variability of extragalactic radio sources, especially the flat spectrum sources, was discovered a long time ago. Typical for such variable sources is the BL Lac object S5 1803+78 (Kühr et al. 1981b) which shows a flat radio spectrum, an optical continuum with no emission lines, strong optical polarization, and variability both in the radio and optical range (Biermann et al. 1981). Its spectrum is shown in Fig. 1. It is especially interesting that for such sources the X-ray luminosity is approximately equal to the mm-wave luminosity (Owen et al. 1981). A contrasting overall spectrum is exhibited by M82 (Fig. 2) which is dominated by the emission from a recent burst of star formation (Rieke et al. 1980, Kronberg et al. 1981, 1985); here the emission in the mm range is due to hot dust emission from clouds heated by young OB stars.

*Dedicated to Prof. R. Kippenhahn on the occasion of his sixtieth birthday

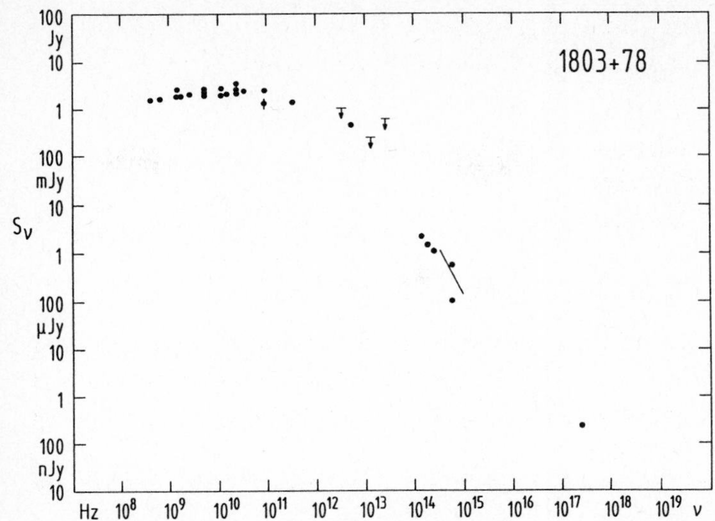

Fig. 1: The total spectrum of the BL Lac object S5 1803+78 (Chini et al. 1986)

The variability of radio sources led M. Rees (1966) to argue that such sources are in a state of relativistic expansion: Synchrotron self-absorption gives a lower limit on the size, while the variability time scales give an upper limit. These two limits are in strong disagreement for a number of sources and so an obvious and simple way out is to consider relativistic bulk motion. This leads to two further important effects: First, emission from a source which moves relativistically ($v \simeq c$) with respect to an observer is boosted in the direction of motion, leading to extreme intensity increases of the radiation in this direction. Second, and even more exciting, is the possibility of observing apparent superluminal motion. This effect arises again most strongly for observers near to the line of motion, and is due to a kinematic illusion caused by the large light travel time differences. Using the Lorentz factor γ to describe the bulk velocity, we find for $\gamma \gg 1$ that $v_{app} \simeq \gamma c$ for the angle between the line of sight to the observer and the direction of motion $\theta \simeq 1/\gamma$.

Very long baseline interferometry (VLBI) discovered apparent superluminal motion in 1971 for the first time (see the review by Kellermann and Pauliny-Toth 1981). Since then this effect has been seen in numerous compact radio sources.

The basic concept used today for the interpretation of compact radio sources (Blandford and Königl 1979) comprises the following ingredients: A compact jet is made up of a series of emission knots. Each knot emits Synchrotron radiation which is self-absorbed at low frequency. The absorption frequency naturally goes up with compactness and so we are looking deeper into the nucleus at higher radio frequencies. The various emission knots together produce in superposition a nearly flat radio spectrum. But there are lots of questions:

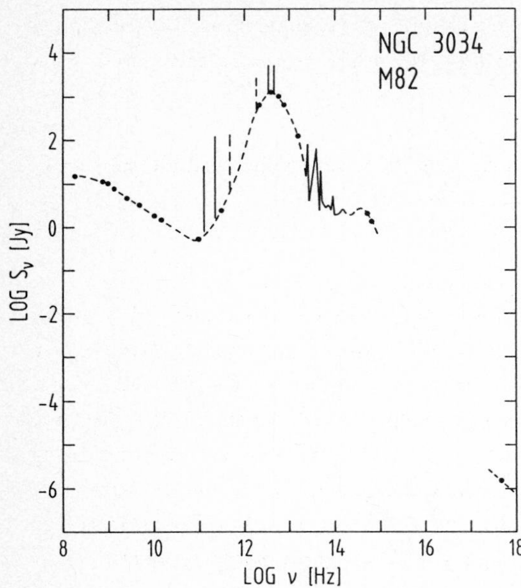

Fig. 2: The total spectrum of the starburst galaxy M82 (Kronberg et al. 1985)

1) Why is activity in nuclei so common?
2) How do we obtain up to ~10^{48}erg/sec in luminosity and have variability?
3) How can jets maintain their coherence from scales of below 1 pc to ~ 1 Mpc despite their variability?
4) How can we explain photons emitted in the energy range from ~ 10^{-7}eV to ~ 10^{9}eV? For two cases (3C273, Cen A) we even know that near photon energies of ~1 MeV the luminosity is higher than at any other observable photon energy (Bezler et al. 1984, Ballmoos 1985).
5) Can active nuclei be the sources of very high energy cosmic rays, of energies above 10^{19}eV?

VLA radio surveys have demonstrated that virtually all bright quasars have radio emission, for optically bright quasars we reach completeness at a level of flux density at 5 GHz of about 100 μJy (Kellermann, priv. comm.). Other radio source surveys have demonstrated that virtually all extended radio sources have compact cores. Naturally, the question arises how many active nuclei show apparent superluminal motion if we accept the simple kinematic model?

It is clear that if the presence of relativistic bulk motion could be demonstrated convincingly for an entire class of active nuclei, then the path to follow in our investigations to understand the physics of active nuclei would be clearer.

Therefore I wish to discuss now two complete samples of radio sources, one designed not to show evidence of bulk relativistic motion, the other to positively show such effects.

Were we to find bulk relativistic motion by detecting apparent superluminal motion in both samples, then our simple concept may have to be abandoned.

THE STEEP SPECTRUM SAMPLE

Over a number of years Porcas and Zensus have investigated the cores of a small number of sources chosen from a complete sample of 30 steep spectrum radio sources (Owen et al. 1978, Porcas 1981, 1982, 1984, 1985a, b, 1986a, b; Zensus and Porcas 1986; Zensus et al. 1986). These sources are selected by flux density, by the property of being a quasar, by having a largest angular size over 10 arc seconds and declination above 40° from the 966 MHz Jodrell Bank survey. Most of these sources are dominated by extended steep spectrum emission. This extended emission arises from two usually symmetric radio lobes (Owen and Puschell 1984). The symmetry, similar flux densities and sizes of the two lobes have been used to argue that the lobes cannot be in bulk relativistic motion (Longair and Riley 1979). Even mild relativistic motion ($\gamma \approx 2$) can have dramatic effects to produce highly asymmetric structures. The two lobes are believed to be fed by jets, often undetected, that define an axis for the sources. If the total emission is dominated by the radiation from extended symmetric lobes, then it appears rather certain that these lobes cannot move at relativistic velocities. If so, it follows that their radiation is emitted isotropically. And as a consequence a sample selected solely by the strength of this emission should be isotropically distributed in its orientation relative to the line of sight to the observer. Even if all sources were to have bulk relativistic motion along the jets feeding the radio lobes with energy, apparent superluminal motion should show up only for a small fraction of the sample. We have

$$\beta_\perp = v_{app}/c = \frac{\beta \sin \theta}{1-\beta \cos \theta} \tag{1}$$

where θ is the angle between the line of motion and the line of sight to the observer, $\beta=v/c$ and v the bulk belocity. We can detect apparent superluminal motion at significant levels usually if $\beta_\perp \geqslant 2$ which imposes a limit on θ by

$$\cos \theta \geqslant \frac{1}{1+\beta_\perp^2} \left(\frac{\beta_\perp^2 \gamma}{\sqrt{\gamma^2-1}} - \sqrt{1 - \frac{\beta_\perp^2}{\gamma^2-1}} \right) \tag{2}$$

where $\gamma^2 = 1/(1-\beta^2)$.

For $\beta_\perp \geqslant 2$ and, e.g., $\gamma=10$ as suggested by many examples of known sources with apparent superluminal motion, we thus obtain $\theta \leqslant 52^\circ.5$. If sources eject matter at bulk relativistic velocities symmetrically to both sides, as suggested by overall symmetry of sources and the asymmetry of the compact sources (due to relativistic boosting of the radiation of one side over the other), then the fraction P of all sources that should show apparent superluminal motion is given by

$$P = 1 - \cos \theta^{*} \qquad (3)$$

where θ^{*} is given by the limit of eq. (2). For the limit of $\beta_{\perp}=2$, and $\gamma=10$ we thus obtain a probability of 39% that we find apparent superluminal motion in a randomly oriented complete sample.

Porcas and Zensus have actually observed 3 sources so far with sufficient data to say something about apparent superluminal motion. All three sources show the effect with 3C179 the best known example . For randomly chosen sources the probability is thus 6%, that all three should show the effect, again assuming $\beta_{\perp}=2$ and $\gamma=10$ to represent the observational limit and a "typical" Lorentz factor. 6% is not entirely negligible, and so one could call it "luck".

However, there is an important selection effect: If, as assumed, all cores have bulk relativistic motion, then their radiation field is boosted (Rybicki and Lightman 1979)

$$\left(\frac{dP}{d\Omega d\nu}\right)_{observer} = \left(\frac{1}{\gamma(1-\beta \cos \theta)}\right)^{3} \left(\frac{dP}{d\Omega d\nu}\right)_{source} \qquad (4)$$

where θ is measured in the frame of the observer. Here β corresponds to the velocity of the frame in which the radiation is isotropic, not generally the same velocity at which we observe apparent superluminal motion. The boosting factor $D = 1/(\gamma(1-\beta \cos \theta))$ is a very strong monotonic function of θ, reaching 2γ for $\cos \theta=1$, $1/2\gamma$ for $\cos \theta=-1$, and is unity at $\sin \theta=\sqrt{2/(\gamma+1)}$, which means $\theta=25.^{\circ}2$ for $\gamma=10$. It follows that in a sample of randomly oriented sources a subselection for sources with sufficiently strong cores to do VLBI is likely to pick out exactly those which are pointing approximately towards the observer.

Porcas and Zensus, on the other hand, tried very hard to avoid just this selection effect which is well understood but is difficult to completely avoid because of the sensitivity limit of VLBI.

Götz and Preuss (Götz 1986) have also made VLBI observations of the cores of two classical "edge-brightened" radio galaxies; again the goal was to test radio sources for evidence of bulk relativistic motion when it is not expected because of the overall symmetry of the source. They find evidence for apparent superluminal motion in one (3C390.3) of these sources, again quite contrary to expectation.

In conclusion, the success of Zensus and Porcas, and Götz and Preuss to find apparent superluminal motion in radio sources for which we do not expect it on the basis of their likely orientation, remains intriguing.

THE FLAT SPECTRUM SAMPLE

A group of scientists around Witzel has worked for some years now on a sample of radio sources selected to test the opposite from the situation discussed above: Using a complete sample of 13 core dominated sources from the 5 GHz MPI-NRAO survey (Kühr et al. 1981a), the S5-survey (Kühr et al. 1981b) ($\delta \geqslant 70°$) with the properties S(5 GHz)$\geqslant$1 Jy, and spectral α index between 2.8 and 5 GHz flatter than −0.5 ($S_\nu \sim \nu^\alpha$), they are conducting an extensive observational program including multiple epoch VLBI observations at a variety of frequencies, and observations at all accessible wavelengths from the radio to the X-ray range.

Observational results and interpretations from this program have been partially published (Biermann et al. 1981, 1982; Eckart et al. 1982, 1985, 1986a, b; Eckart and Witzel 1983; Johnston et al. 1984, 1986; Witzel et al. 1986).

For this sample the canonical model for compact jets (Blandford and Königl 1979) is generally believed to be applicable: A compact relativistic jet is pointing close to the line sight to the observer, and the superposition of the individual, partially self-absorbed synchrotron emission spectra from the individual knots produce a flat radio spectrum. Consequently these sources should all be variable, should have a strong inverse Compton X-ray flux, should often show apparent superluminal motion. We have been able to verify all these expectations.

These sources are all thus found to show evidence for bulk relativistic motion with $\gamma \simeq 10$ (Eckart et al. 1986a, b; Schalinski et al. 1986).

A noteworthy example is the quasar S5 1928+738 (Eckart et al. 1985, Johnston et al. 1986): At redshift z=0.302 (Lawrence et al. 1986) it shows a long apparently superluminal jet with five knots moving outwards observed to 17 milli arc seconds at 5 GHz. Using the canonical model again, we can deproject the length of the jet. We find a length for that part of the jet which shows evidence of bulk relativistic motion of several hundred parsecs. Assuming that the outer radio structure is collinear with the compact structure, this leads to a deprojected overall size of about 1 Mpc, making S5 1928+738 one of the largest radio sources known. Even if the angle between compact and extended structure were large (see Johnston et al. 1986), S5 1928+738 would remain a very large source of a size well over 200 kpc (H_0=100 km sec^{-1}Mpc^{-1}, q_0=0.05). This leads to the obvious question of where the bulk relativistic motion becomes subrelativistic. Such a transition requires an enormous amount of dissipation:

To go from $\gamma=10$ to, say, $\beta=0.1$ reduces the kinetic energy of the flow by a factor of nearly 2000, while to go from $\beta=0.1$ to $\beta=0.01$, the reduction in kinetic energy is "only" obviously a factor of 100. The transition to subrelativistic velocities can thus be expected to lead to strong observational effects, visible as very strong emission. In a source such as S5 1928+738 there is indeed an emission knot which stands out, the outermost one on the same side where apparent superluminal motion is visible. We are thus led to the speculation that relativistic velocities may be maintained to the outermost knot and thus lead to relativistic boosting of the radiation from the knot itself, not from the extended emission around it.

However, if this picture should be correct then we are forced to reconsider the case of the steep spectrum sample. For sources like 3C179 and 3C390.3 there is also a strong emission knot visible at the outer edge of the radio lobe on the same side as where the apparent superluminal motion is visible. It may a common feature that apparently superluminal jets point at the more compact hot spot (Linfield 1981, and see the remark of Porcas (1983) at the Turin meeting). If the radiation from this knot is relativistically boosted then we have a small second selection effect in the steep spectrum sample: Selected by strength of emission from the outer lobes, a small part of which may be boosted, we in fact bias our sample to select boosted sources: Remembering again that the subsample selected for VLBI observations was selected partially on core strength because of VLBI sensitivity limits, it should come as no surprise that the cores show apparent superluminal motion.

CONCLUSION

Consideration of both a steep spectrum sample and a flat spectrum sample of radio sources and the detection of evidence for bulk relativistic motion in practically all sources of both samples which have been carefully studied, leads us to the following hypothesis: In many radio sources the bulk relativistic motion persists out to the radio lobes.

ACKNOWLEDGEMENTS

Intense discussions with Drs. R.W. Porcas and A. Witzel are gratefully acknowledged.

REFERENCES

Ballmoos, P.V.: 1985, Ph.D. Thesis, MPE Report 193

Bezler, M., Kendziorra, E., Staubert, R., Hasinger, G., Pietsch, W., Reppin, C., Trümper, J. Voges, W.: 1984, Astron. Astrophys. 136, 351

Biermann, P., Dürbeck, H., Eckart, A., Fricke, K., Johnston, K.J., Kühr, H., Liebert, J., Pauliny-Toth, I.I.K., Schleicher, H., Stockman, H., Strittmatter, P.A., Witzel, A.: 1981, Astrophys. J. Lett. 247, L53

Biermann, P., Fricke, K., Johnston, K.J., Kühr, H., Pauliny-Toth, I.I.K., Strittmatter, P.A., Urbanik, M., Witzel, A.: 1982, Astrophys. J. Lett. 252, L1

Blandford, R.D., Königl, A.: 1979, Astrophys. J. 232, 34

Chini, R., Biermann, P.L., Kreysa, E., Kühr, H., Mezger, P.G., Schmidt, J., Witzel, A.,

Zensus, J.A.: 1986, in preparation
Eckart, A., Hill, P., Johnston, K.J., Pauliny-Toth, I.I.K., Spencer, J.H., Witzel, A.: 1982, Astron. Astrophys. 108, 157
Eckart, A., Witzel, A.: 1983, IAU Symp. No. 110, 65
Eckart, A., Witzel, A., Biermann, P., Pearson, T.J., Readhead, A.C.S., Johnston, K.J.: 1985, Astrophys. J. Lett. 296, L23
Eckart, A., Witzel, A., Biermann, P., Johnston, K.J., Simon, R., Schalinski, C., Kühr, H.: 1986a, Astron. Astrophys. (in press)
Eckart, A., Witzel, A., Biermann, P., Johnston, K.J., Simon, R., Schalinski, C., Kühr, H.: 1986b, Astron. Astrophys. Suppl. (in press)
Götz, M.: 1986, Univ. Bonn, Diplom-Arbeit
Johnston, K.J., Biermann, P., Eckart, A., Kühr, H., Strittmatter, P., Strom, R.G., Witzel, A., Zensus, A.: 1984, Astrophys. J. 280, 542
Johnston, K.J., Simon, R.S.A., Eckart, A., Biermann, P., Schalinski, C., Witzel, A., Strom, R.G.: 1986, Astrophys. J. Lett. (in press)
Kellermann, K.I., Pauliny-Toth, I.I.K.: 1981, Ann. Rev. Astron. Astrophys. 19, 373
Kronberg, P.P., Biermann, P., Schwab, F.R.: 1981, Astrophys. J. 246, 751
Kronberg, P.P., Biermann, P., Schwab, F.R.: 1985, Astrophys. J. 291, 693
Kühr, H., Witzel, A., Pauliny-Toth, I.I.K., Nauber, U.: 1981a, Astron. Astrophys. Suppl. 45, 367
Kühr, H., Pauliny-Toth, I.I.K., Witzel, A., Schmidt, J.: 1981b, Astron. J. 86, 854
Lawrence, C.R., Pearson, T.J., Readhead, A.C.S., Unwin, S.C.: 1986, Astron. J. 91, 494
Lind, K.R., Blandford, R.D.: 1985, Astrophys. J. 295, 358
Linfield, R.: 1981, Astrophys. J. 244, 436
Longair, M.S., Riley, J.M.: 1979, Monthly Notices Roy. Astron. Soc. 188, 625
Norman, M.L., Winkler, K.-H.A., Smarr, L.L.: 1984, Proc. "Physics of Energy Transport in Extragalactic Radio Sources", eds. A. Bridle, J. Eilek
Owen, F.N., Porcas, R.W., Neff, S.G.: 1978, Astron. J. 83, 1009
Owen, F.N., Helfand, D.J., Spangler, S.R.: 1981, Astrophys. J. 250, L55
Owen, F.N., Puschell, J.J.: 1984, Astron. J. 89, 932
Porcas, R.W.: 1981, Nature 294, 47
Porcas, R.W.: 1982, IAU Symposium 97, 361
Porcas, R.W.: 1983, Turin meeting "Astrophysical Jets", eds. A. Ferrari, A.G. Pacholczyk, 36
Porcas, R.W.: 1984, IAU Symposium No. 110, 157
Porcas, R.W.: 1985a, "Extragalactic Energetic Sources", ed. V.K. Kapahi, Indian Academy of Sciences, Bangalore, 113
Porcas, R.W.: 1985b, Manchester Conference on Active Galactic Nuclei, ed. Dyson, Manchester, 20
Porcas, R.W.: 1986a, Mitt. Astron. Ges. 65, 95
Porcas, R.W.: 1986b, IAU Symposium No. 119, 131
Rees, M.J.: 1966, Nature 211, 468
Rieke, G.H., Lebofsky, M.J., Thompson, R.I., Low, F.J., Tokunaga, A.T.: 1980, Astrophys. J. 238, 24
Rybicki, G.B., Lightman, A.P.: 1979, "Radiative Processes in Astrophysics", Wiley-Interscience, New York
Schalinski, C., Witzel, A., Biermann, P., Eckart, A., Johnston, K.J.: 1986, Mitt. Astron. Ges. 65, 244
Witzel, A., Eckart, A., Biermann, P., Schalinski, C., Johnston, K.J.: 1986, Mitt. Astron. Ges. 65, 242
Zensus, J.A., Porcas, R.W.: 1986, IAU Symposium 119, 167
Zensus, J.A., Hough, D.H., Porcas, R.W.: 1986, Nature (in press)

Optical Emission from Radio Hotspots

H.-J. Röser *

Max-Planck-Institut für Astronomie, Königstuhl 17,
D-6900 Heidelberg 1, Fed. Rep. of Germany

I Introduction

Radio hotspots - knots of high surface brightness in the outskirts of the lobes in classical double radio sources - are naturally explained as the working surface of a jet, supposed to supply energy, mass and momentum from the nucleus to the lobes (see e.g. Begelmann et al. 1984). Their study should thus provide important clues to our understanding of the structure and physics of radio galaxies. The advent of large aperture synthesis radio telescopes as MERLIN and the VLA has provided beautiful maps of these sources with an overwhelming amount of detail on the arcsec to subarcsecond scale (Heeschen, Wade 1982). Interpretation of these structures is, however, severely hampered by the lack of velocity information and the fact that the relativistic particles giving rise to the observed radio synchrotron radiation can travel an appreciable distance from their place of birth (synchrotron lifetime ~ 10^4 years in a magnetic field of 1 mG and for an observational frequency of 1 GHz). The situation is completely different for the optical wavelength range, where the lifetime of the particles is on the order of only 100 years, i.e. we do observe places of particle acceleration directly (e.g. shock front structures, field reconnection points etc.).

Extrapolating a typical hotspot continuum spectrum (e.g. Cygnus A; Wright, Birkinshaw 1984) from radio to optical frequencies, one expects optical counterparts to be readily observable on the Palomar Sky Survey prints. As this is generally not the case, a spectral cutoff or at least a turnover has to occur somewhere between the radio and optical frequency range. According to canonical synchrotron theory, such a cutoff could be directly related to the maximum energy the radiating particles can acquire (Pacholczyk 1970, 1977). This would certainly be an important quantity to measure in order to provide constraints for our theories of particle acceleration mechanisms in these sources.

---

* This work was done in close collaboration with Dr. Klaus Meisenheimer and, recently, also Dr. Peter Hiltner.

Nevertheless very little has been done in the past regarding optical identification of radio hotspot counterparts. Available data in the literature are either limited mainly to deep direct imaging (Simkin 1978, Crane et al. 1983, Kronberg et al. 1977) or are concerned with line emission, which as turned out, does originate not from the jet or hotspot themselves but from their vicinity (van Breugel 1986 and references therein). Although necessary for a start, the first approach obviously is liable to chance coincidences by faint unrelated objects and - more severe - the fact that radio and optical brightness maxima do not necessarily coincide (Röser, Meisenheimer 1986). We therefore decided to use a different approach and chose field polarimetry of all objects in a CCD frame centered on the hotspots radio position to pick out polarized objects related to the hotspot (for details of the technique see Meisenheimer, Röser 1986b).

II Results

In the following, a brief summary of our previous work is given.

a) 3C 33 south : Results of our optical observations of 3C 33 south and their interpretation have been published elsewhere (Meisenheimer, Röser 1986a). To summarize, we detected high, linear polarization in the optical counterpart reported by Simkin (1978) in a red bandpass, closely matching radio polarization properties as reported by Rudnick et al. (1980). Whereas on Arp's deep 200" III-aJ plate used by Simkin hardly any radiation seems to originate from the hotspot itself, recent CCD observations at the prime focus of the 3.5m telescope on Calar Alto clearly detected the hotspot counterpart also in a blue (λ = 423nm) and a near infrared (λ = 860nm) bandpass. Based on our R band polarimetry and the B direct image (the I band data are not yet reduced), the continuum shape exhibits a synchrotron cutoff spectrum (Pacholczyk 1970, 1977) with a cutoff frequency of $1.3x10^{14}$ Hz. Such a spectrum was also found for the hotspot in 3C 273A (Röser, Meisenheimer 1986) with a similar cutoff frequency. We regard these observations as the first direct evidence of optical synchrotron radiation from a hotspot in a classical double radio source. If we estimate the magnetic field from equipartition considerations, Lorentz factors on the order of $\gamma = 10^5$ are required for the radiating particles.

b) Pictor A (west) : The area of the western hotspot in the double radio source Pictor A (0518-456) was observed in November 1985 with EFOSC at the 3.6m telescope on La Silla (Röser, Meisenheimer 1987). Again polarization measurement was accomplished by means of a Savart plate in the diverging beam. We detected a highly polarized (>30%) object of B = 19 mag within 3" of the radio position (beam ~10"x25") of the hotspot (Prestage 1985). The object is also marginally detected in an EINSTEIN IPC frame. The radio-to-optical spectral index of -0.9 steepens slightly to -1.0 for the optical-to-x-ray index. This is in marked contrast to the cutoff spectra in 3C33s and 3C273A. At the moment, we cannot proof the synchrotron origin of the x-ray radiation, although an inverse Compton or Bremsstrahlung origin does seem unlikely. Should it, however, turn out to be of synchrotron origin, particles with $\gamma = 10^7$ would be required.

III Current and future observational efforts

To shed more light onto the physical background leading to vastly different values of the cutoff frequency, i.e. the maximum energy of the radiating particles and/or the value of the magnetic field strengths, observations of more hotspots are needed. We therefore concentrate our efforts on the well defined sample of 166 3CR sources, observed at 5 GHz with the Cambridge interferometer at 1" to 2" resolution (Jenkins et al. 1977). A statistical analysis of the hotspot's 5 GHz brightness in a 1" aperture (relevant for optical observations) was used to select the most promising objects for our direct imaging and polarimetry using CCD detectors in the prime focus of the 3.5m telescope on Calar Alto. We have to keep in mind, however, that with respect to optical detectability, a significant shift in cutoff frequency may well be more important than the absolute 5 GHz flux density.

For a few selected sources, first preliminary results can already be given :

3C 6.1 : several faint objects were found on a 1500 sec exposure (seeing 1.6") close to the southern hotspot location (radio - optical position differ by less than 2.5"). Due to the moderately dense field ($bII = 17^0$) , its relationship to the hotspot radiation has to be proven by polarimetry.

3C 20 : This source also is at low galactic latitude ($bII=-11^0$). We found a faint object coinciding with the northern hotspot location within the astrometric error of ~0.7". Polarization data have been collected but not yet analyzed.

3C 86A : This is a new identifcation of a low galactic latitude source with a galaxy. An object 2" off the hotspot location does not show any high polarization and seems to be an unrelated foreground object.

3C 111 : A very faint object in an otherwise scarcely populated field is coinciding with the eastern hotspot. Polarimetry is not yet available.

3C 123 : A very deep CCD frame (R filter, exposure time 2000 sec at 1.4" seeing) does not show any counterpart of the hotspot.

3C 319 : The object reported by Crane et al. (1983) and barely visible on the POSS E plate seems to be a cluster galaxy on the bases of its morphology on a much deeper frame than that presented by Crane et al. . There is also no high polarization detected.

3C 327 : The object visible on both POSS plates seems to be a galaxy according to its appearance on a deep prime focus Savart exposure.

3C 330 : On a deeper image than the one presented by Crane et al. we also do see radiation originating from the hotspot vicinity, but at the moment cannot proof its positional coincidence nor its polarization properties.

3C 405 (Cygnus A) : Due to the very crowded field we have problems with our CCD polarimetry. We do have excellent polarization data, but we lack at the moment a deep direct image necessary to properly reduce them.

3C 454.3 : Deep direct B and R images taken at the ESO 3.5m telescope with EFOSC do not reveal any counterpart of the jet and hotspot.

IV Summary

Detection of optical (and possible x-ray) radiation from radio hotspots in classical double radio sources and verifying its synchrotron origin by polarization data, enabled us to derive the maximum energy of the radiating particles. For a magnetic field strength on the order of 1 mG we derive Lorentz factors of 10^5 for 3C273A and 3C33s. In the case of Pictor A, where we may also see the hotspot at x-ray frequencies, γ may be as high as 10^7. These observations provide an important data base also for a detailed mapping of the regions of particle acceleration by the Hubble Space Telescope. Detailed observations of these and other hotspots are in progress.

References

Begelman,M.C., Blandford,R.D., Rees,M.J. : 1984, Rev.Mod.Phys. 56, 255

Crane,P., Tyson,J.A., Saslaw,W.C. : 1983, Astrophys.J. 265, 681

Heeschen,D., Wade, C.M. (editors) : 1982, IAU Symposium 97 "Extragalactic Radiosources", Reidel, Dordrecht

Jenkins,C.J., Pooley,G.G., Riley,J.M. : 1977, Mem.R.A.S. 84, 61

Kronberg,P., van den Bergh,S., Button,S. : Astron.J. 82,315

Meisenheimer,K., Röser,H.-J. : 1986a, Nature 319, 459

Meisenheimer,K., Röser,H.-J. : 1986b, proceedings of the ESO/OHP workshop on CCDs

Pacholczyk, A.G. : 1970, Radio Astrophysics, Freeman, San Francisco

Pacholczyk, A.G. : 1977, Radio Galaxies, Pergamon Press, Oxford

Prestage,R.M. : 1985, Ph.D. thesis, University of Edinburgh

Röser,H.-J., Meisenheimer,K. : 1986, Astron.Astrophys. 154, 15

Röser,H.-J., Meisenheimer,K. : 1987, Astrophys.J. 314, in press

Rudnick,L., Saslaw,W.C., Crane,P., Tyson,J.A. : 1981, Astrophys.J. 246, 647

Simkin, S.M. : 1978, Astrophys.J. Lett. 222, L55

van den Breugel,W. : 1986, Can.J.Phys. 64, 392

Wright,M., Birkinshaw,M. : 1984, Astrophys.J. 281, 135

Index of Contributors

Springer